UNIFORMS OF THE REPUBLIC OF TEXAS

And the Men that Wore Them

1836-1846

Bruce Marshall

Schiffer Military History
Atglen, PA

Book Design by Robert Biondi.

Library of Congress Control Number: 98-86680

ISBN: 978-0-7643-0682-2
Printed in China

Schiffer Books are available at special discounts for bulk purchases for sales promotions or premiums. Special editions, including personalized covers, corporate imprints, and excerpts can be created in large quantities for special needs. For more information contact the publisher:

Published by Schiffer Publishing Ltd.
4880 Lower Valley Road
Atglen, PA 19310
Phone: (610) 593-1777; Fax: (610) 593-2002
E-mail: Info@schifferbooks.com

For the largest selection of fine reference books on this and related subjects, please visit our website at **www.schifferbooks.com**
We are always looking for people to write books on new and related subjects. If you have an idea for a book, please contact us at proposals@schifferbooks.com

This book may be purchased from the publisher.
Include $5.00 for shipping.
Please try your bookstore first.
You may write for a free catalog.

In Europe, Schiffer books are distributed by
Bushwood Books
6 Marksbury Ave.
Kew Gardens
Surrey TW9 4JF England
Phone: 44 (0) 20 8392 8585; Fax: 44 (0) 20 8392 9876
E-mail: info@bushwoodbooks.co.uk
Website: www.bushwoodbooks.co.uk

FOREWORD

We who have served in the military forces of Texas are grateful to Bruce Marshall for his contribution to the preservation of our heritage.

Only the state of Texas possesses a military heritage that includes the army and navy of an independent nation.

Bruce Marshall's paintings portray history in a manner that reflects the great pride of those who have served in this unique military force.

Thomas S. Bishop
Major General, (Ret.)
Army of the United States

INTRODUCTION

Enter with me now into a long-forgotten vault of history to rediscover the uniforms of the Republic of Texas, lost for over a century and a half. Most historians, even in Texas, have been unaware of them. Those I have consulted, for the most part, have expressed the skepticism that: *there weren't any uniforms*, or, *if there were they were only drawing board creations*. This book will dispel all doubt that they were ordered, inspected, paid for, shipped and received, issued, worn and, when worn out, auctioned.

All this is in the brittle, yellowing documents in the archives, libraries and collections I have researched, reinforced by actual contemporary descriptions by those who saw them and, in some cases, wore them. Still preserved, though some not where you might expect to find them, are the contracts for the uniforms and accoutrements, certificates of inspection, correspondence showing when they were shipped and even on what vessels, Quartermaster General inventories for the army and militia, Navy Yard and pursers records of the various ships of the Texas Navy, giving the exact amount in inventory quarterly: how many were issued and how many returned. There were more than even I expected to find. *In fact, at one time the republic had more uniforms than soldiers!*

Guidance in my search, and clues to the locations of certain otherwise unfindable items were provided by others dedicated to Texana. Foremost among these was the late Joseph Hefter, an old Austrian soldier who spent the last and most productive part of his life in Mexico recreating with his meticulous watercolors the uniforms of Texas, Mexico, Spain in the new world, and the Imperium of Maximilian. Others deserving special acknowledgment include: Dr. Tuffly Ellis, former director of the Texas State Historical Association; the late Major General Thomas S. Bishop, the only Adjutant General of the Texas National Guard to serve under two separate governors; novelist Garland Roark, whose specialty was the Texas Navy; Dr. John M. Thiel, Chief of Naval Operations of the Texas Navy and Billy Bob Crim, President of the Texas Navy

Association, who encouraged me with the especially created title *Artist in Residence of the Texas Navy*; former state representative Bill Blythe Jr. and former Speaker of the Texas House of Representatives Bill Clayton for recognizing my work with the specially created title *Artist, Sixty-fifth Legislature*; artist-historian Tom Jones; Texana specialist Colonel John R. Elting of the Company of Military Historians; the late Randy Steffen, who sponsored my membership in the Company of Military Historians and who, in my opinion, was the most accurate of the military artist-historians; the National Guard Association of Texas, which demonstrated its confidence in the accuracy of my work by considering no other artists when commissioning the epic oil *The Texas Citizen Soldier*, which depicts the history of the earliest Texas militia to the present, as their gift for the Texas Room of the National Infantry Museum at Fort Benning, Georgia; Michael J. Koury; publisher of the Old Army Press; and lastly, the most important mentor of my career, ever generous with his time, advise and encouragement, Brigadier General Jay A. Matthews Jr., head of Presidial Press and editor and publisher of *Military History of Texas and the Southwest* (and now editor emeritus of its successor, *Military History of the West*) who, like Hefter, has made his most important contribution to military history after retirement.

Bruce Marshall, K.C.W.E., O.S.J.
Eanes-Marshall Ranch
Austin, Texas

This is Clark L. Owen in his Republic of Texas army uniform, according to his descendent Roberta Adele Brackenridge Menger Byrd. Owen enlisted in the Texas Army as a private in 1836. Later he served as a lieutenant, captain and finally colonel. His service included the Plum Creek Fight against the Indians and the Mier Expedition against the Mexicans. As far as can be determined, this is the only known photo of a uniformed soldier of the Republic of Texas.

A highly retouched photograph of Edwin W. Moore in uniform. But is it his uniform as a lieutenant in the United States Navy before his Texas service? Or his uniform as Commodore of the Texas Navy? The evidence is contradictory, but suggests the latter. An American lieutenant would have an epaulet on his right shoulder *only*. While Moore's uniform does not conform to the Texas regulations of 1839 it is exactly the same as full dress for an American commodore of the same period, with the only difference being a star and *eagle* on the American epaulets and a star and *anchor* on the Texas version (which can't be discerned in the picture). If he is wearing his Texas uniform this is the only known photo of a Texas Navy uniform.

1836

After a seven-month revolution in which the Texians, as they then called themselves, won only the first and last battles, the independence of Texas from Mexico was achieved by the spectacular victory at San Jacinto, April 21, 1836. An election followed in September and by vote of the people a constitution was approved and the first president to serve under it, Sam Houston, elected. With his inauguration, October 22, began the Republic of Texas.

Houston was, first and foremost, a believer in militia as the first line of defense for the new nation. Considering the economic ruin his retreat across Texas had caused during the Texas Revolution, with its scorched earth policy, he was undoubtedly right. However, most agree his negativism towards a navy was shortsighted considering the lengthy coastline of Texas. The militia of Texas was not to be taken lightly, despite their "wild, rough" appearance. As one boasted, "we feel as if we could whip our weight in wildcats and five times our weight in Mexicans." According to a British consul in Texas, they were "the most formidable militia, for their numbers, in the world."[1]

Following the victory at San Jacinto volunteers continued to arrive from the United States, swelling the regular army, which had been only 783 or so men in that final battle. Soon between 2,000 and 2,500 adventurers were on hand and more expected. By this time the Texas Army contained more newly arrived Americans than original Texians.

Such correspondence of the period as is preserved in the Texas State Archives is filled with complaints of inadequate clothing. Yet some had arrived during the closing months of the revolutionary provisional government of interim President David G. Burnet. About a week after San Jacinto Commissary General John Forbes purchased from Colonel James Powers: 100 pair duck pantaloons; 365 cotton shirts; and 14 (illegible). During July the new Texas agent in New Orleans, Thomas Toby and Brother, shipped to Texas: 500 cartridge boxes; a like number of bayonet belts and scabbards; 500 black leather caps and haversacks; two cases containing twelve dozen striped twilled pants; two dozen duck twilled pants; two dozen duck twilled shirts; twenty-four dozen white cotton pants; 500 pint tin cups; 596 soldiers canteens. In September Toby forwarded: seventy canteens; eight dozen tin cups; 400 pair of brogans; eleven and a half dozen white wool hats; twenty dozen Russia duck shirts; ten dozen brown linen pantaloons; twenty dozen blue twilled pantaloons; seventeen dozen brown jackets; 303 best red flannel shirts. Also to come, Toby advised, were: forty dozen duck shirts; five dozen brown linen pantaloons; twenty-five dozen blue drilling pantaloons; 297 red flannel shirts; and 200 pair of brogans. In addition, Commissary General Forbes had on order: 1,000 caps; 1,000 leather stocks; 1,000 cotton duck roundabouts; 1,500 cotton duck pantaloons; 1,000 pair of socks; 2,000 pair of shoes assorted; 1,000 haversacks; 1,000 knapsacks; and other military supplies.[2]

In the same order Forbes asked for "300 yards of gray cloth for officers uniforms and gold lace for trimming."[3] From this it would appear that, while looking to the United States as its primary source of military clothing, Texas was preparing to make some of its own. The provisional government, established by the convention which met from March 1 through 17, had made provisions for "2,000 gray suits" as uniforms for the army.[4] But there is no documentary evidence that, if ordered, they were paid for and delivered. Texas, during the revolution and republic, had difficulty obtaining credit. More than once, items weren't received because of the Texians' inability to pay for them. For example, in 1839, 1,500 Model 1816 flintlock muskets were ordered from Tryon Son and Company of Philadelphia, Pennsylvania. Eight-hundred and sixty were paid for and received. The balance of 640 were sent by Tryon to a warehouse in New Orleans to be delivered upon payment. They were still languishing there when the Republic of Texas ceased to exist in 1846.[5]

Forbes' effort to standardize Texas Army uniforms, at least those for officers, according to what was prescribed by the provisional government in March would have to find fulfillment later, however. The clothing obtained for the military by Texas agents in New Orleans and elsewhere during the interim presidency of Burnet was, for the most part, anything they could lay their hands upon on short notice and even shorter credit. Caps were black leather, sealskin or Japanned. Roundabout jackets and pants were usually white, sometimes brown. Shoes were normally black, sometimes brown or kip. Cartridge boxes were black with white straps. Belts were black. These were usually surplus American Army or militia issue. Plate 1 reconstructs what most of the soldiers must have looked like during the early months of the republic.

★ ★ ★

Plate 1 (page 33)
REPUBLIC OF TEXAS ARMY
Fall 1836

Most of the soldiers outfitted by the government looked like this. The majority wore black caps of the hunting style, but some wool hats were also issued. So the soldier is shown trying on a black hunting cap, but holding one of the wool hats. Both his roundabout jacket and pantaloons are white. The collar of his jacket is the citizens style of the period rather than the military stand-up collar adopted later. At this time much, perhaps most, of the clothing sent by Texas agents in New Orleans and elsewhere was whatever they could get their hands on because of the haste necessary. Efforts were made to Texanize and militarize it by substitution of Texas buttons, as shown by the correspondence of interim President Burnet and others, and perhaps the addition of lone star insignia, probably cut out of tin or cast from brass.

In this plate and the following two the banner in the background is the Zavala flag. Designed by interim Vice President Lorenzo de Zavala, it was adopted in May of 1836 by the interim revolutionary government. It served the republic from October to early December, when it was replaced by another designed by David G. Burnet, known as the Burnet flag.[6]

★ ★ ★

Plate 2 shows some of the variations that were also worn, or possibly worn, during the fall of 1836.

★ ★ ★

Plate 2 (page 34)
TEXAS ARMY
One of several possible early variations

In August and September, during the closing days of the interim regime, the Texas agent in New Orleans, Thomas Toby and Brother, shipped a considerable amount of supplies for the Texas Army. Included were brown jackets, pants and shoes, plus 500 black leather caps. Undoubtedly these supplies were used by the republic. Most probably the brown and kip brogans were worn with the brown jackets and pants. All of the cartridge boxes were black, but the straps were most likely white U.S. Army surplus, as shown here and in Plate 1. Some might have been black. The best evidence is that the caps were the popular hunting style shown in Plate 1. But some contend the caps may have been the U.S. military model of 1833, nicknamed the "hogkiller," shown here.

Plate 1 shows the roundabout jacket as *single* breasted, whereas in this plate it is *double* breasted. Roundabouts of the period came in both styles, and with some variation of collar. But the Texas records do not reveal exactly which types were received by the Texas Army during this period.

★ ★ ★

During the republic the army shifted camp five times. All of these camps were within a ten-mile radius along certain waterways. Relic hunters have found numerous Texas Army buttons at these sites. This is conclusive proof that the clothing being received was Texanized by addition of military buttons at the very least. The buttons for the Texas army and navy came mainly from Scovill of Waterbury, Connecticut. Scovill records show purchases beginning in September of 1836.[7] Some buttons may also have been purchased earlier from other sources.

The government continued its recruiting campaign in the United States, though it could ill-afford its growing army. According to the final report of outgoing president Burnet, delivered the same day Houston took office, the national debt was $1.25 million and the treasury was almost empty.[8] Furthermore, the army was troublesome. The soldiers were hard to control because many units had arrived with their own officers unamenable to submission to the regular Texas command. They were restless for action, unhappy with the republic's inability to pay them. Citizens complained of possessions and livestock being confiscated by the newcomers. Mexican-Texans, *Tejanos* as they called themselves, were a particular target of the late arrivals. Animated by racial loyalty and outrage at the many acts of barbaric cruelty and treachery by Santa Anna and his minions, the late arrivals were eager to settle accounts with *any* Mexican. Schemes began to hatch among the idle soldiers for reckless forays into Mexico to discipline the perpetrators of the massacres at the Alamo and Goliad, and gain recompense while at it.

Heading this volatile assemblage was the most reckless adventurer of the lot, Felix Huston. As observed by a traveler from Virginia, Colonel William Fairfax Gray, he was outfitted "*a la militaire*."[9] At the time his rank was brevet brigadier general.

★ ★ ★

Plate 3 (page 35)
FELIX HUSTON
Brigadier General

Felix Huston became Commander-in-Chief of the Texas Army, succeeding Thomas Jefferson Rusk, President Houston's first appointment to the command. Near year's end, Houston appointed Albert Sidney Johnston as Senior Brigadier General and Huston as Junior Brigadier General. This made Johnston commander. But when he attempted to assume command in early 1837 he was challenged to a duel by Huston. After an exchange of six shots Johnston was seriously wounded. Huston thus retained command for the time being by this unorthodox means.[10] Later their positions were reversed and Johnston was commander with Huston serving under him, their differences apparently resolved.

Still later, in November of 1839, Huston won election to the position of Major General of the Militia over San Jacinto veteran Sidney Sherman.[11] He served against the Comanches in the 1840s.

Throughout his involvement with the Texas military Huston was a firebrand, ceaselessly agitating for invasion of Mexico. He represented the more aggressive element of the military adventurers attracted to Texas during the revolution and the republic.

Described as "popular and capable"[12] but also as "turbulent and overbearing,"[13] Huston apparently had no military experience prior to joining the Texas Army. He was a native of Kentucky who migrated to Mississippi before settling in Texas. Later in life, his ambitious plans for Texas expansion frustrated, he returned to the United States.

Huston had a waist-length portrait painted wearing the uniform shown in Plate 3.[14] This would be an undress uniform rather than full-dress because it is single-breasted. But at what point in his career the portrait was done is not known. Thus this might be a self-designed uniform made when he was busy in the United States recruiting during the revolution. If so, he undoubtedly wore it to Texas when he arrived sometime after San Jacinto. It might also be of later vintage, but before 1839. The 1839 regulations, which are very detailed, designate double-breasted coats for both full-dress and undress. And the 1839 buttons for a brigadier general are spaced in pairs of two each, not as shown in Huston's portrait. There are other differences as well that clearly make this a pre-1839 uniform. The soft-top garrison cap, while not in his portrait, would be appropriate for a general in undress uniform either during the revolution or the republic. The sash for a brigadier general was red (yellow for major general).

A unique peculiarity of Texas uniforms was that such sashes were to be tied on the right hip. In almost every other army of the world the sash is tied on the left side. Texans have always enjoyed being different.

★ ★ ★

Among the first formal uniforms authorized to be worn for certain was that of the dragoons. On the surface this elaborate and gaudy uniform for these mounted infantry would appear to be highly unlikely, especially considering the utilitarian modesty of those previously shown. However, there is concrete evidence that it was worn.

★ ★ ★

Plate 4 (page 36)
TEXAS DRAGOONS
1836

This elegant uniform had a leather jockey cap with plume that was undoubtedly copied from that worn by the United States dragoons. It is mentioned in the diaries of General Thomas Jefferson Chambers, who may have ordered them – at least a prototype – from a maker in Kansas City, Missouri. The tailcoat was a brilliant green with gold collar, epaulets and frogs. Trousers were an egg-shell white or buff color, probably nankeen. High black jackboots and brass spurs completed the uniform. We know exactly what it looked like, even the dragoon horse equipment, from a sketch made by William T. Rainey, who served in the Texas Dragoons in 1836. The original sketch is in the Corcoran Gallery in Washington, D.C. Rainey's service and his sketch are conclusive proof this uniform existed, was worn, and was not just a drawing board model.[15]

The late Joseph Hefter did two color studies of this uniform but in one the spurs shown are white metal. Brass spurs are shown here as only brass ones are mentioned in purchases or later quartermaster inventories or regulations. Hefter says the uniform in the sketch is that of a "subaltern officer in one of the early units of the Texas Army of the Reserve." The design of this uniform can probably be credited to Chambers, who was a major general of the Army of the Reserve and, as his correspondence has shown, authorized procurement or manufacture of certain uniforms, some possibly from his own designs. His papers concerning uniforms frequently list "tunics of lindsey" and there is mention of "green Marion cloth for uniforms." This could well be the cloth for the dragoon coat, especially as there appear to have been no other green uniforms for the army. There is also mention of "dies for helmets ... mounting for helmet ... striking up helmet and valise."[16] These almost certainly were for the dragoon helmet.

In the background is the Burnet flag, designed by David G. Burnet. It replaced the Zavala flag in early December of 1836. It remained the Texas national flag until 1839, when replaced by the lone star banner which served until the end of the republic. The lone star flag has continued to the present day as the Texas State flag.[17]

★ ★ ★

As 1836 drew to a close the government had a novel problem with uniforms – counterfeits. A number of unauthorized persons posing as Texas officers were strutting about in uniforms and insignia of the Texas Army, shaking down the gullible for contributions in the United States and in some cases confiscating property in Texas in the name of the government. On December 19, Acting Secretary of War William S. Fisher felt it necessary to issue a General Order warning about such unauthorized persons. "The impostors," he fumed, "by their licentious and unprincipled conduct, and many impositions upon the friends of Texas abroad, reflect disgrace on the country and service, which they pretend to represent."[18]

1837

Commissary of Purchase records reveal that during 1837 Colonel A.S. Thurston, Commissary General of Texas, journeyed to New Orleans and perhaps elsewhere, making the following purchases:

February 7 – 208 seal(skin) caps; 880 pair men's kip brogans; 2 copper bugles; 1 doz. c. fifes; 5 ebony fifes, silver mounted; bugle instructions fife and music.[19]
April 5 – 1,093 forage caps, bought from John Heavy.[20]
May 11 – 424 cartridge boxes and belts, from John Y. Coey (or Hoey).[21] Also on the same date: 19 muskets with bayonets were received from Abraham Miller and paid for with government bonds.[22]
June-August – 424 cartridge boxes from John Hoey. It is not clear if this represents delivery of the 424 cartridge boxes bought on May 11, or if it is a new order. On the document is a notation by Hoey, "received for the above five sections of Texas land."[23]

Not shown in the archival records, but mentioned in the recollections of Noah Smithwick, was the purchase in 1837 of some surplus U.S. Army uniforms. Some of these were sent to the Texas Rangers stationed at Coleman's Fort between Austin and Bastrop. Smithwick was one of the rangers receiving the uniforms. This blows up the long-held belief of the rangers and their admirers that the rangers never wore uniforms. It is hard to understand how this cherished myth has held on to the present day since Smithwick's book, *The Evolution of a State or Recollections of Old Texas Days*, was published in 1900 and reprinted in 1968.

★ ★ ★

Plate 5 (page 37)
TEXAS RANGERS
Coleman's Fort – 1837

Until the American uniforms arrived the Rangers at Coleman's Fort had worn their own civilian clothing, mostly buckskins, according to Smithwick. The U.S. clothing consisted of "pantaloons and runabouts," not complete uniforms. What Smithwick calls a "runabout" was actually a *roundabout*, often called in the records of the archives a *round jacket*. It was popular with both army and navy uniforms as well as with civilians and common seamen. It is also known as a *shell jacket* in military parlance. It was waist length. The army version usually had a stand-up collar, whereas the naval and civilian versions had a collar similar to the World War II era "Ike" jacket.

Most of the uniforms sent to the fort were "rather under sized," according to Smithwick. Adjustments, some on the comical side, were made by some of the recipients or their wives in order for them to squeeze into them. This produced some interesting variations, Smithwick remembered:

> "Isaac Casner, who tipped the beam at 200, got a suit that would have been snug for a man of 140. As the old fellow couldn't begin to get into them he took them on his arm and went round among the boys trying to effect an exchange. We all liked Uncle Isaac and the largest suits in the lot were brought out. He tried them on one after another, but like the 'Contraband's Song,' 'they wouldn't go half way around,' and but for the ingenuity of Mrs. Casner the old man's suit would have been a total loss. Clothes were scarce, though, so Mrs. Casner ripped open the outside seams of the pantaloons and set in stripes to extend them to the necessary dimensions, also setting a stripe down each sleeve and in the center of the jacket, with a false front to expand it over his aldermanic proportions. A stranger would have taken him for commanding officer on account of his stripes.
>
> "Walfenberger, who would have measured six feet barefoot got a suit of which the bottoms of the pantaloons struck him about half way to his knees, the jacket failing to connect with them by full six inches, and his arms protruding a foot beyond the end of his sleeves. He presented a ludicrous appearance as, amid shouts of laughter, he stalked up and down like an animated scarecrow, trying to negotiate a trade. Failing in that, he pieced them out with strips of blanket and was quite as comfortable as the rest of us."[24]

Since the uniforms didn't include headgear or footwear, these are shown as being what they ordinarily would have worn, a wide brimmed hat and moccasins, which were often decorated with beading Indian-style.

The ranger carries his long arm in a fringed leather sheath to protect it and keep his powder dry. He carries a Spanish gourd canteen. These gourds held about two quarts.

Coleman's Fort, also known as Fort Coleman, is shown in the background as Smithwick sketched it.

★ ★ ★

Correspondence between Texas agent B. Hammatt Norton and others and the Scovill company in the Texas Button File of the Baker Library, Harvard Business School, documents that the purchase of military buttons began in 1836 continued into 1837. There is additional evidence that Scovill produced other buttons for Texas even later, at least until 1840. Still other buttons were supplied from France, the markings on the backs showing a coat of arms and G & Cie, Paris or V•SUPERFIN•G•PERFECTIONNE. Both are believed to have been made by Garanti, a Parisian manufacturer. Another button of French manufacture is marked on the back T.W.&W. Paris. This was undoubtedly Trelon, Weldon and Weil, Paris, a supplier of buttons worldwide.[25]

1838

Another indication that the republic was having at least some of its military uniforms made in Texas is a purchase June 19, 1838 by Commissary General A.L. Huston from S.W. Tibbulls of clothing materials. Tibbulls received $10,080. On the same date Colonel Sidney Sherman was paid $2,930 for "superintending and having clothes made" using the materials from Tibbulls.[26]

In August four boxes of "public clothing" were shipped from Cincinnati, Ohio to New Orleans, Louisiana via Pierce Shannon & Brother Freighting, as authorized by A.S. Thurston.[27]

Verifying that uniforms were issued and worn is a form listing return of clothes issued to cavalry recruits at the Port of Galveston from September 1 through October 22: 17 jackets; 19 pantaloons; shirts (none); socks (none); 17 pair shoes; 17 caps; 34 blankets.[28]

The archival records are by no means complete. Some were lost by fire or other mishap. In addition to uniforms issued to the enlisted personnel, officers often purchased their own uniforms. This was probably the case with Juan Seguín, who had his portrait painted by Thomas Jefferson Wright in 1838 wearing his blue dress uniform (field uniforms were gray) as a lieutenant colonel of Texas cavalry.

Juan Seguín was the son of Erasmo Seguín, an alcalde of San Antonio and former member of the Mexican Congress. The aristocratic Seguíns had welcomed the original Anglo *empressario* Moses Austin to Texas at the beginning of Anglo colonization. From then and throughout the revolution they sided with the colonists and Mexican liberals against the central government, especially following the discarding of the constitution of 1824. Juan had been at the Alamo, but was sent out by Travis with one of Travis' several appeals for assistance. Thus he escaped the fate of the other Alamo defenders. He commanded a company of *Tejanos* at San Jacinto. After San Jacinto, he became mayor of San Antonio and about the same time received his commission as lieutenant colonel of cavalry in the regular army.

★ ★ ★

Plate 6 (page 38)

LIEUTENANT COLONEL JUAN SEGUÍN
Cavalry – 1838

The uniforms of the republic were in the main influenced by American uniforms of the period, and to a lesser extent by contemporary Mexican uniforms. The Mexican uniforms, in turn, were influenced by their Spanish predecessors and to some degree by Napoleonic fashions. The most obvious difference between Texan and American uniforms was the buttons, of course, and the frequent use of lone star insignia. But as Texas matured the changes increased. Seguín's uniform is not totally in harmony with the Texas Army regulations promulgated the following year. The 1839 regulations were probably meant to bring into uniformity existing dress, rather than create new uniforms per se. Seguín's uniform is an example of why this conformity was desirable. For instance, it was a particularity of Texas officer sashes that they be tied on the *right* side, but

Seguín's sash, though not showing completely in the waist-length portrait, appears to be tied on the *left*.[29] His helmet is virtually the same leather jockey cap of the dragoon model, prescribed by General Chambers, with its white plume. Seguín's helmet has an additional small tuft of red plume, which was authorized for a colonel. The sunburst insignia has a lone star replacing the eagle of the American version of the helmet. While Seguín's trousers don't show in his portrait, they would be the same dark blue as his coat for winter, with a buff stripe (the cavalry color) down the seams. For summer the pants would be white linen without the stripe. Cavalry officers wore boots. Enlisted men wore bootees.[30]

★ ★ ★

During a visit to Houston, then the capital of Texas, Colonel Edward Stiff observed and described one of the best uniformed Texas Army units, the Milam Guards. They were the bodyguard of the president, then Sam Houston. Houston for most of his life was a profane, hard-drinking womanizer. His escort reflected the chief executive's appetites and vices. So did the city named after him for that matter. Stiff called it a "great rendezvous for abandoned characters from the four corners of the globe."[31] Another traveler, Francis C. Sheridan, described it as "the most uncivilized place in Texas."[32]

The Milam Guards wore the infantry dress uniform which was the same as shown in Plate 7 and Plate 8, as prescribed by the regulations of 1839. Headwear was a black shako with a white plume. Insignia on it was a sunburst with a lone star like that on the dragoon jockey helmet. Coat and trousers were dark blue in winter with white duck linen pants for summer. Infantry had white metal buttons and white trim. Footwear was black boots for officers, booties for enlisted ranks.[33]

Both the president and his bodyguard were inebriated when Stiff saw them, not surprisingly:

> "The president entered the town escorted by the Milam Guards, whose white pantaloons were in strange contrast with the torrents of rain descending, and the half-leg deep mud in the streets, which at a short distance gave each man the appearance of a pair of black boots drawn over his inexpressibles and the illusion might have been complete had not a shoe been occasionally lost in the mud (which caused the heroes to halt until the barefoot man could recover his understanding)."

Stiff uses the words boot and shoe interchangeably. And the words were used interchangeably in those days, and still are to some extent today in the British isles. But the footwear mentioned was undoubtedly the bootee, a high-topped military shoe. The balance of Stiff's observation of the Milam Guards is worth repeating:

> "Arriving at the White House ... the guards entered, and stacked their arms on the porch between a brace of which the president entered followed by as many thirsty and hungry beings as was ever congregated in the most rarefied society.
>
> "For two days this revelry was kept up, amid the beating of drums, firing of guns, cutting of throats and a confusion of tongues ... On the second night the guards escorted the president to the theater ... A peal of three cheers proclaimed the arrival of the president and suit which was speedily followed by a hissing, the discharge of pistols, the glittering of Bowie knives, while many a knight proclaimed his prowess by a volley of profanity, some leveled at the president, some at the mayor, some at the police; when at length all seemed exhausted, the field of battle was examined and three reported wounded; killed none."[34]

The perils faced by the Texas military were apparently not all from Mexicans and Indians.

As 1838 wound down, Mirabeau Buonaparte Lamar took over as President. His increasing hostility towards Sam Houston was no doubt aggravated by Houston's pompous performance at the inauguration on December 1. Houston upstaged Lamar with a three-hour outgoing harangue of self-praise and comparison of himself to George Washington while outfitted in a colonial costume and powdered wig. Lamar was so disgusted he handed his speech to his secretary to read for him and went home.[35]

Immediately Lamar reversed most of Houston's policies, *particularly military*. Whatever his devious political shifts, Houston had always been at heart an annexation proponent. Lamar wanted an independent nation, envisioning it expanding westward to the Pacific coast, perhaps even rivaling the United States. This required a big military. The shift was now from reliance on militia in a defensive posture to a regular army and navy which would take the offensive against the Mexicans and Indians.

Even before the end of the year an elite unit was created, the First Regiment of Infantry. It was to be smartly uniformed and armed with the finest of weaponry. Texas agents would soon be on their way to the United States to obtain the uniforms and equipment that would be required. During the reign of Lamar the army and navy of Texas flowered in all the glory the republic's limited resources would permit, and beyond.

1839

Now began the important years for Texas uniforms. During 1839 exact and highly detailed regulations, long overdue, were promulgated for both army and navy.

Those for the army were published under date of May 23 as General Order Number 5, *Uniform of the Army of the Republic of Texas*. It was issued from the office of the Adjutant General, Albert Sidney Johnston, at Houston by order of the Secretary of War, Hugh McLeod. The preface read:

> "The President has been pleased to adopt the present uniform for the army of the Republic of Texas. It will be the duty of the officers and men to adhere rigidly to it."[36]

Uniforms had long existed, as mentioned. These regulations merely attempted to standardize them. To an extent they did, though not totally. Individualism, lack of the designated materials and varying interpretations still presented problems. All services and ranks, except common seamen, marines and militia, were covered and dress and materials meticulously detailed in minutiae. Even patterns of the cut were furnished or models of the headwear. Unfortunately none of these patterns or models have survived for us to examine. Throughout the regulations runs the instruction "according to pattern." But the patterns are missing. They may have been lost in certain fires that destroyed much of the military records. However, from reading the contracts entered into by Texas agents with American manufacturers, it may be surmised that the patterns and models were left in the hands of the manufacturers. It would seem strange, however, that duplicate patterns were not kept by the government for comparison with the finished product when delivered. Again, this puzzlement may be explained by the several Certificates of Inspection at the point of manufacture in the United States or nearby. These inspections were done by United States government military inspectors moonlighting for the Republic of Texas. Duplicate patterns must have been left with them also. In cases where there was no pattern, the inspectors gave the assurance that the items were made equal to their counterpart in U.S. Army clothing or equipment.

The late Joseph Hefter did a credible job in reconstructing what the cut of Texas uniforms must have been, guiding on the cut of contemporary American uniforms, after which most of the Texas uniforms were styled but also mindful of Mexican, Spanish, Napoleonic and even British influences. The new regulations continued blue as the color for dress uniforms and gray for field service, with white pants for summer and, for the cavalry at least, continued the all-white summer fatigue uniform. Appearing perhaps for the first time were red jackets for cavalry buglers.

Plates 7 through 21 show the uniforms of the Republic of Texas as prescribed by the regulations of 1839, and as modified by actual usage. This same year the lone star flag was adopted, superseding all other Texas flags (though certain of the naval flags continued to be used unofficially for the deceptive advantage they afforded).

★ ★ ★

Plate 7 (page 39)

INFANTRY

Captain – Full Dress

This is the dress-uniform of a Texas officer. We know from the observations of Colonel Stiff that the infantry dress uniform was worn by at least one unit, the Milam Guards, even before the regulations of 1839.

The officer's white plume is more elaborate than the tuft topping the enlisted shako. Each shako had the sunburst insignia with a lone star in its center. Centered in the star was a letter showing the wearer's company. Below the sunburst was a numeral signifying his regiment. Collar trim was silver lace. A captain had two "loops" on each side ending at a button. A lieutenant had a single loop on each side plus the button. A captain had two white leg stripes, a lieutenant one. A captain had one epaulet on his right shoulder, a lieutenant one on his left shoulder. Each wore a scale on the other shoulder. Sashes were crimson and tied on the right. The aguillettes were for commissioned staff only, and of twisted cord. Boots were Wellingtons.[37]

The lone star flag behind was adopted the same year as the uniform regulations.

★ ★ ★

Plate 8 (page 40)

INFANTRY

Sergeant – Milam Guards

Dress Uniform

The shako of the enlisted men was identical to that of the officers except for a tuft of white instead of the feathered plume, and it lacked the tassels of silver cord and bullion. Collar trim was white on the blue uniform, not silver like on the officer shako. Shoulder scales and leg stripes were also white. Sergeants wore their chevrons on the right sleeve only, above the elbow. Corporals wore theirs also on the right sleeve only, but below the elbow. Non-commissioned staff and first sergeants wore a red sash. White trousers were worn in summer. The dress uniform of other infantry was no different from that worn by the Milam Guards.[38]

The lone star belt plate shown in this illustration and the following one is not mentioned in the regulations but can be found in the Cressman contract of 1839.[39]

The First regiment was equipped with Tryon flintlock muskets, Model of 1812.[40]

★ ★ ★

Plate 9 (page 41)

INFANTRY

Fatigue Uniform

This was the uniform most used by the Texas infantry. The First Regiment wore it, or the summer version, in its campaign against the Cherokee Indians, when garrisoning the various frontier forts and camps on the frontier, and while building the military road that was to link various points within the republic. The fatigue uniform was worn by regulars also during the battle with the Comanche Indians at Plum Creek and possibly at the Council House fight. Color was gray, trim was black. Buttons for infantry were white metal. Depicted is a first sergeant, distinguished by the straight bar under his chevrons. As with the dress blues, only noncommissioned staff and first sergeants wore the red sash. Garrison caps for all branches of the army were dark blue, but sometimes the infantry wore on theirs a black silk oilcloth cover for weatherproofing. A gray greatcoat with cape top and stand-up collar was prescribed and a few turn up in the inventory lists.[41]

This soldier is shown with a tin canteen. Canteens of both tin and wood were bought by Texas agents or manufactured for the army. A considerable number of pint tin cups were also purchased, plus haversacks, knapsacks, belts, cartridge boxes and belts for them, bayonets and scabbards. Except for the Texas plates (of white metal for infantry) on the buckles, boxes and their belts, these were made to U.S. Army specifications.

To the naked eye the only apparent difference between the fatigue uniforms for winter and summer was a change in summer to pants of white duck. However, the most important difference was the material of the summer jacket. The winter one, and the trousers that went with it, was of kersey cloth (a blend of wool with linen or cotton). The summer jacket of the Texas uniform was cotton, however, and it probably was a shell jacket rather than the tailcoat worn in winter. Thus the Texas summer uniform, unlike that of the American army, was cool and practical. The U.S. uniform, both top and bottom, winter or summer, was kersey. This was true throughout the remainder of the nineteenth century in American uniforms and also with most Confederate uniforms as well. Enlisted men in the infantry, as well as other branches of the army, wore black booties. These may have been, like their American counterpart, capable of fitting either foot.[42]

There may also have been an all white fatigue uniform for summer, with the jacket very plain and without any collar trim.

★ ★ ★

Plate 10 (page 42)

THOMAS JEFFERSON CHAMBERS
Major General

Thomas Jefferson Chambers received the rank of Major General in the Army of the Reserve from the interim government during the revolution. It was bestowed on the understanding that Chambers would raise, arm, and uniform emigrant volunteers from the United States. He did so, but neither he nor his recruits arrived in time for the fighting. His role as Major General of the Army of the Reserve continued into the republic. Pictured in this plate is the uniform he had made for himself. It was in most respects identical to the prescribed regulation one of 1839 with only a couple of variations. This is another reminder that the 1839 regulations did not create new uniforms, but were intended merely to standardize already existing uniforms. His bicorn hat is worn sideways here, but it could be worn fore-and-aft; *sometimes,* in both the American and foreign armies, especially during the Napoleonic period, even at a rakish sideways angle. The epaulets of Chambers' uniform had three stars, whereas the regulations of 1839 call for a single star on the epaulet for a general of any rank. The three stars signified major general in the Texas Army. They were worn on a shoulder strap on the undress uniform. A Texas brigadier general wore two stars. This was different from the American army, where a major general had two stars and a brigadier general had one. The other difference between Chambers' uniform and the 1839 regulations was his gold belt, instead of a black one.[43]

There seemed to have been only two ranks of general in the Texas Army, major general and brigadier general. The dress uniform was the same except for buttons and sash. Each rank wore ten gold buttons, but the higher rank had them spaced by threes with a single one at the top on each side as shown here. The brigadier general had his spaced in pairs. A major general wore a gold sash, a brigadier general a red one. What is left of Chambers' uniform and regalia is on public display at the San Jacinto Monument near Houston: one epaulet, the belt and sword shown, plus a martingale ornament from his horse equipment. The only other Texas uniform pieces extant, Sidney Sherman's coat and vest (worn during the revolution) are also at the San Jacinto Monument.[44]

Rumor from a well-placed source has it that the rest of Chambers' uniform, in remarkably good condition, is also in the San Jacinto Monument, but kept in storage. In response to an inquiry about this, the museum's reply curiously didn't confirm or deny the existence of the rest of the uniform. It merely confirmed that the components mentioned above, as well as Sherman's coat and vest, were on public display.[45]

Like Felix Huston, General Chambers served under all of the flags of Texas. Shown here is the lone star banner.

★ ★ ★

While the First Regiment of Infantry was the elite unit of the Texas Army before it was disbanded in 1841, the cavalry played in many ways a more important role. This was particularly true because of the vast distances to be covered and the mobility of the Indians, who were usually mounted. The designation *dragoons* seems to have been used interchangeably with *cavalry* by the Texas military. Most armies use the designation *dragoons* to mean mounted infantry. The Texas cavalry/dragoons had quite a few stylish uniforms. All were worn.

★ ★ ★

Plate 11 (page 43)

CAVALRY
Sergeant – Full Dress
the Travis Guards

This is a sergeant of cavalry in full dress uniform. At least one company was outfitted in the dress uniform, the Travis Guards. This we know from a requisition in the Texas State Archives from the captain of the Travis Guards, Alex Woodhouse, for cavalry uniforms for thirty-three men plus two musicians coats. At the bottom of the requisition is a notation by Captain Woodhouse that the uniforms had been received.[46]

A peculiarity of this single-breasted coat is that the center has nine buttons, whereas there are eleven on each side. Another unusual feature is that the skirt of the coat in only four inches long. Non-commissioned staff wore yellow bullion on the scale on the left shoulder and an ordinary scale on the right. Other enlisted men wore regular scales on each shoulder. Headwear was the plumed leather jockey cap.

Cavalry trim and buttons were gold. Sergeants wore their chevrons on both sleeves above the elbow, point downward. Corporals had two bars below the elbow, point downward; lance-corporals one bar.[47]

According to some scholars, at least one unit had a red uniform. This uniform was requested by Captain A.S. Plummer in a letter to President Lamar dated April 5, 1839. "Give me a fancy uniform of red," Plummer enthused, "and I can enlist first-rate men – nothing like a dashing uniform ..." According to Mike Koury of the Old Army Press, Plummer got the red uniforms, which were to be for dress.[48] But the only red clothing that shows up in the inventories is red jackets for musicians.

The trouser stripe and sword knot are buff, which was the cavalry color (not yellow). The trouper's neckwear is a black leather stock.

The Travis Guards were actually volunteer militia, not regular cavalry. They were at Austin. When the capital was moved to Austin they served as an honor guard for the president, as the Milam Guards had done in Houston. On these occasions they wore the dress uniform shown. When in the field they no doubt wore the fatigue uniform.

The horses in the background are wearing the Texas Army cavalry saddle, bridle, martingale and pistol holsters, of which there was one on each side of the pommel. Texas Army horses were branded "TA" on the right shoulder.[49]

The small, squarish flag in the background is the Texas cavalry standard, prescribed by the regulations of 1839.[50]

★ ★ ★

Plate 12 (page 44)
CAVALRY
Trumpeter

A variation of the full dress uniform was red jackets for musicians (these were usually trumpeters or buglers). All else, including the bob-tailed cut of the jacket, was the same as for other cavalry troopers. These red jackets were definitely worn because they show up regularly in the inventories. The distinctive red musicians jacket was also a feature of the U.S. cavalry of the period.[51]

Like the troopers in Plate 11 and Plate 14, the trumpeter carries an Ames sabre in a browned scabbard. Two hundred eighty of these sabres were made for Texas by N.P. Ames of Springfield, Massachusetts, under a contract signed by Texas agent William Henry Daingerfield, February 4, 1840. They were the U.S. Model 1833. The only difference between the ones used by the U.S. cavalry and the Texians was that the Texas version had the words "Texas Dragoons" engraved on the blades and a lone star on the hilts.[52]

The designation 1st Cavalry on the flag indicates the First Texas Cavalry Regiment. This regiment was never up to full strength, however. Its actual size was probably only about a battalion (between one hundred twenty to four hundred men).[53]

★ ★ ★

Plate 13 (page 45)
CAVALRY
Officer – Summer Undress Uniform

This uniform, both winter (with blue pants) and summer (with white pants) was worn extensively. All of the military personnel on the Santa Fe expedition was uniformed records of the expedition reveal. The numerous requisitions indicate that most, perhaps all of the pioneers (as they were called), received the undress uniform with both the blue *and* white pants, plus the gray fatigue uniform *and* the all-white summer fatigues. Presumably they meant to make an impressive appearance on arrival and expected to be gone a long time. This officer wears the blue cloth fatigue cap. His stock is black silk or black leather. The tail coat is cut on civilian lines with a "falling collar," rather than the usual standup collar. This type of *falling collar* in recent times has been called a *stand-and-fall* collar.[54]

★ ★ ★

Plate 14 (page 46)
CAVALRY
Corporal – Fatigue Dress

The gray fatigue uniform was also used extensively. Like the infantry version, trim was black, the winter version was kersey cloth, the summer jacket cotton and the summer pants white duck. Cavalry buttons and buckles were gold, rather than the infantry white metal. The sword knot and strap were buff colored.[55]

Behind is the flag of the First Cavalry Regiment.

★ ★ ★

Plate 15 (page 47)
CAVALRY
Summer Fatigue Uniform

The cavalry regulations of 1839 also called for white cotton drilling for summer. These were on the same pattern as the gray fatigue uniform except much plainer. Many are shown in the quarterly inventories. This uniform was undoubtedly copied from the U.S. Army summer whites of the same time frame.[56]

The trooper wears one of the brown cotton shirts, purchased in addition to the white ones used more frequently. His stock is black leather. The blue cap seems to have been used with all undress and fatigue uniforms.

Shown in addition to the cavalry regimental flag is a cavalry guidon. Each company carried one with its initial centered in the star. The cavalry flag was two feet six inches wide and two feet four inches high. The guidon was three feet from the lance to the end of the slit of the swallow-tail, and two feet high. The lance for both flag and guidon was nine feet tall, including the spear and ferule.[57]

★ ★ ★

Two months earlier, comprehensive uniform regulations for the navy were issued by Naval General Order, dated March 13. Like those for the army, naval uniforms already existed; these regulations merely formalized them. Navy buttons, for instance, were among the Scovill orders of 1836 and 1837. While not as detailed as the army regulations, those for the navy were explicit as to what the officers were to wear, and also gunners, boatswains and sailmakers. They are silent as to the clothing for common seamen, but their winter and summer uniforms, curiously enough, were the only uniforms detailed in the Regulations and Instructions for the Government of the Naval Service of Texas of December 15, 1836.[58]

Contrary to the belief of many historians and most Texas Navy buffs, Texas Navy officers uniforms were styled primarily after those of the British navy, according to the research of novelist Garland Roark, author of *Star in the Rigging* about the Texas Navy. American fashions were only a secondary influence. This should not be a surprise. While the United States was the parent country of most Texians, the British navy then, and long into the twentieth century, was considered the foremost navy of the world. Traditionally military fashions are copied after the most successful in war.

The patterns for the Texas Navy uniforms, like those for the army, have unfortunately disappeared. But their cut may be surmised by reference to those of the British and American uniforms of the period.

The first Texas Navy of the revolutionary period and into the first years of the republic prior to 1839 had by now lost all of its ships. New ones were purchased beginning in 1839. These were principally the steamship *Zavala*, the schooners *San Jacinto*, *San Antonio* and *San Bernard*, the brig *Wharton* and the sloop-of-war *Austin*, the flagship. This was the second Texas Navy. The uniform regulations of 1839 are therefore particularly associated with it. The same year the lone star flag was adopted, superseding all other Texas flags including those of the Navy.

Under Commodore Moore the navy became the pride of Texas and the principal deterrent to a re-invasion by Mexico, despite many efforts at sabotage by Sam Houston during his second presidential term.

★ ★ ★

Plate 16 (page 48)
CAPTAIN – NAVY
(and Commodore)
Full Dress

Captain was the highest official rank in the navy. But the senior captain of the second navy, Edwin W. Moore, was always called Commodore. The uniform for captain and Commodore was one and the same.

The full dress uniforms had the tailcoat shown but two variations in trousers. The one shown is the more formal, with nearly obsolete knee-breeches. Plain white pantaloons over "short boots" were an alternate. The pantaloons could also be blue over "half boots" (probably the same as "short boots").[59]

There were two versions of the undress uniform. The first should perhaps be described here as it was identical to the full dress uniform except for the elimination of the oak leaf trim, and thus really an only slightly modified version of full dress.[60]

Commanders wore the same formal dress as captains, except for the oak leaves, the pennant embroidered on the collar was smaller, and they had only three buttons on cuffs and under pocket flaps instead of the four for captain.[61]

Shown behind is the Hawkins naval flag, though it had supposedly been retired in 1839. The Hawkins flag and the variant flag (a cross between the Hawkins and lone star flags) continued to be used because of their deceptive resemblance to the American flag. This was helpful in confusing Mexican ships. Still further deception with misleading flags is suggested by the 1841 quarterly returns of Moore's flagship, the *Austin*. On board were ensigns and pennants of the United States, Britain, France and even Mexico. Apparently Moore placed some value on the saying "All is fair in love and war," at least the latter part.[62]

★ ★ ★

Plate 17 (page 49)
NAVY – CAPTAIN
Summer Undress Uniform

As mentioned in the description in Plate 16, there was an undress uniform that was virtually the same as the full dress uniform. The other version, shown here, was probably the uniform worn most when at sea: a double-breasted frock coat with standing collar. Epaulets as in the full dress version were regulation, but the more modest shoulder tabs shown in this plate were probably more commonly worn. In winter blue trousers were worn.[63]

Most of the few Texas naval swords preserved appear to be ones the officers purchased personally as there is little conformity of style and they are from several makers. A contemporary description of the sword of Commodore Moore says it was an elaborately decorated one with a big lone star on the hilt.

The captain holds a Patterson Colt revolver in his left hand. The navy was the first of the Texas forces to use the Patterson, which was the first Colt revolver. Virtually all histories of the Texas Rangers claim the Patterson was introduced into Texas by the Rangers, and first used in a skirmish with Indians at Enchanted Rock. This is false. "The Colt's pistols used by the Texas Rangers before annexation, were all supplied from the navy, after they had been in constant use in that arm of the service upwards of four years," Commodore Moore told the Military Affairs Committee of the United States Senate.[64]

Another cherished Ranger myth bites the dust in these pages. Other correspondence reveals that the Commodore, as further directed, also turned over some of the navy's Pattersons to the Santa Fe Pioneers. Some of those Colts were captured by the Mexicans, while others were destroyed by the Texians rather than let such prized weaponry fall into the hands of the Mexicans.[65]

★ ★ ★

Plate 18 (page 50)
NAVY – MIDSHIPMAN

This midshipman wears the summer undress uniform. For winter the pants would be blue and he would wear a blue garrison cap.[66]

His straw hat bears special notice. Sailor straws were traditionally black, usually waterproofed with tar or whatever. While Texas sailors wore the traditional black straw in winter, they also wore these unusual summer straws, apparently peculiarly Texian. According to Garland Roark and other sources, the summer straws were made by the seamen themselves from "hands of grass."[67] This is confirmed by the navy inventories which show many "hands of grass" aboard. These straws were a hybrid between the traditional sailor straw and a Mexican sombrero. Whereas the regular black straw had a more or less straight brim, the Texas summer straw, which was natural straw color, had a brim that curled up hiding some or most of the crown. All straws had a black or navy blue ribbon with a single tassel, according to Roark. Hefter shows two tassels in his art works.

In addition to the Hawkins flag, the ship pennant of the *Wharton* is shown behind the midshipman.

The long hair shown here and in Plate 21 was apparently not uncommon among Texas Naval officers if we are to believe Francis C. Sheridan, who gave an unflattering description of them in his journal of a visit to Galveston and the Texas coast during 1839-1840. Sheridan found:

> "The schooners are pretty vessels but I can't say the same for the others neither can I for the *officers of the Texas Navy* generally speaking. These *take a delight, after the effeminate fashion of the French, in allowing the hair to grow down the back*, which of all the damnable fooleries ever introduced is the most damnable. It is neither cleanly or becoming, and is infinitely more ridiculous than if they were to turn it up behind and stick a large tortoise shell comb with gold knobs on it, after the manner of women. But if it is bad when there is an artificial or natural curl, what is it when there is no curl at all and when *their hair hangs down like the matted ends of a wet swab – and this is generally the case with these officers.*"[68]

★ ★ ★

Plate 19 (page 51)
NAVY – GUNNER

The blue jacket and bell-bottom trousers worn by the gunner shown here were the uniform of all seamen, though the gunners and other specialists had certain insignia to distinguish them, such as the buttons on the gunner's lapels. He wears the traditional sailor straw of most navies of that period. The untanned leather scabbard holds a sheath knife. Red vests were to be worn in winter.[69]

Behind the seaman is the pennant of the steamship *Zavala*. The pennant for the schooner *San Jacinto* was apparently identical to that of the *Zavala*.[70]

★ ★ ★

Plate 20 (page 52)
NAVY – SEAMAN
Summer Whites

This was the summer uniform for seamen. The hand made straw hat is the same as worn by the midshipman in Plate 18. According to Hefter and others, a peculiarity of Texas Navy uniforms was a red scarf. They seem to be basing this on a certain packet of documents in the Mexican archives discovered by Hefter. They describe a raid on the Mexican coastal town of Dizlam (Silam or Silan in Texas documents) in 1837 by the *Invincible* and *Brutus*, ships of the first Texas Navy. These first-hand accounts were in 1976 translated and published in a bulletin by the Rosenberg Library of Galveston. While these are indeed interesting eyewitness descriptions of the ships, flag, and weaponry of the Texians, several readings of the translation fail to locate any description of the Texian uniforms. So where actual documentation of the red scarves can be found cannot be determined. This is not to say Hefter was wrong. There might be still other documentation in the Mexican archives he was aware of, but unknown to those who follow his work. It is certain they wore black kerchiefs, for many black silk handkerchiefs are mentioned in the inventories of the Navy Yard and the vessels of the fleet. For this reason the kerchiefs shown in Plate 19 and Plate 20 are black. Vests of white duck were worn in summer.[71]

For want of a better designation, I call the banner in the background the variant flag. As mentioned earlier, it is a blend of the Hawkins and lone star flags. It was the least used by the Texas Navy. About the only mention of it is in the Dizlan documents. The depositions of two eyewitnesses describe it. Francisco Sosa: "They went up to the forest edge to fasten their flag which was green at its beginning and the end (was) some white and red stripes with a star in the green." Apolinario Lisama: "Said flag was green near the staff side with a white star and on the remainder were red and white stripes."[72]

Hefter shows this flag in one of his uniform plates on the *second* navy with the field next to the staff green, as described by the Mexicans. But it is not likely that any Texas Navy flag had such a green field. A footnote in the Rosenberg Library translation offers the most plausible explanation. "This was probably a Texas Navy flag or ensign with the blue faded to an aquamarine or blue-green by the salt spray."[73]

While officers were entitled to wear swords, aboard ship the favorite of officers and men was a "basket-hilted cut and thrust" not to exceed thirty inches in length nor to be less than twenty-six inches, and to be slightly curved. While regulations have it under swords it appears to better fit the description of a cutlass.[74]

★ ★ ★

Plate 21 (page 53)
NAVY
Foul Weather Clothing

In foul weather everyone aboard the ships wore this informal but practical clothing, with minor variations.

The cap was a black southwester, a waterproof hat made of material such as oilskin or canvas with a broad brim behind to protect the neck. These tarpaulin hats were a very traditional seafarers necessity. They are still worn today all over the world. They appear in the inventories, along with pea jackets such as the one shown. The pea jackets are also shown in the requisitations for the marine guards on the ships. A contemporary sketch by midshipman Edward Johns of the Texas Navy shows both the southwester cap and the pea jacket. His sketch also shows another person in a long overcoat reaching almost to his ankles, possibly an officer.[75] While brass buttons are shown here, it is possible the buttons may have been of horn or bone. The Return of Slops and Small Stores of the *Austin* for the year ending December 13, 1840, show dozens of such buttons in stock, as well as regular gold (brass) ones.[76]

Red flannel shirts show up in the naval inventories as well as white cotton shirts. No doubt the flannel shirts were for cold and inclement weather.

This person, whether officer or sailor, also has the long hair so detested by the traveler Sheridan.

The ship ensign behind is that of the flagship *Austin*. Illustrations for an article on the Texas Navy in the *Republic of Texas* by *American West* magazine and the Texas State Historical Association, as well as the designs in the archives, show all of the ship pennants as being triangular. But a sketch by a Texas Navy officer and a contemporary watercolor show the pennant on the *Austin* as being swallow-tailed, as shown here. Perhaps this was a variation permitted only for the flagship.[77]

★ ★ ★

The Texas Marine Corps had uniforms, though there were no formal regulations governing them. According to Hefter there were three main uniforms, all United States surplus, Texanized with Texas Marine Corps buttons and insignia. Two were formal ones that may have been for shore duty or ceremonial occasions, particularly when recruiting in American ports. The naval inventories suggest that aboard the ships, however, the enlisted marines were more often than not outfitted similarly to the seamen. There is mention, for instance, of the following common clothing in inventories as being identifiably for the marines: duck shirts; duck trousers; flannel shirts; duck frocks; shoes and brogans. Other clothing that was strictly marine in inventories or shown among purchases by Commodore Moore included: green marine caps; marine dress jackets; gray cassinette pants; marine gray jackets; marine white jackets; marine white pants; and marine belts. Raising questions are the marine *gray* jackets and marine *white* pants. These are not depicted by Hefter in any of his watercolor studies, though he was apparently aware of them. The gray fatigue jackets and white pants must have differed in some manner from those for sailors, otherwise they would not be shown as for marines. This difference, at least for the jackets, conceivably could have been only in the buttons or other minutiae. Did the gray pants go with an all gray uniform? Or were the gray pants part of the surplus U.S. uniform of 1834 (which had a green coat and pants that were gray or green, depending upon which authority you are listening to). Almost certainly the gray jackets were to be worn with either gray or white pants, or both. Were they similar to the army gray fatigues, but with marine buttons? Giving some support to this possibility is that some army overcoats, caps (both dress and fatigue), coats, white cotton pants, nankeen pants, and stocks (leather collars) are found in some quarterly returns of the Navy Yard at Galveston. What kind of cap went with the all-white marine uniforms, or the gray one? Blue or green? Both were carried in naval inventories.

★ ★ ★

Plate 22 (page 54)

MARINE LIEUTENANT

Both Hefter and Kenneth L. Smith-Christmas, Curator of Material History, Museums Branch, Marine Corps History and Museums Division, are in agreement that this is an accurate depiction of the U.S. Marine uniform of 1826 (except for the Texas insignia). According to Hefter, a number of these uniforms, and those in the following two plates were "ceded by the U.S. Navy Department to Texas procurement agents, or sold by civilian dealers and shipped to Texas ports in small quantities by merchant vessels."[78] Texas buttons and insignia quickly replaced the American ones. While mentioned in Texas correspondence with button makers, only a handful of Texas Marine Corps buttons have been located. The most recent discoveries were a coat button at the site of Quintana by Cris Kneupper of the Brazosport Archaeological Society, and earlier two others, coat and vest, at Baytown by other relic hunters.

In the background is the variant flag.

★ ★ ★

Plate 23 (page 55)

MARINE CORPS

Sergeant – Fatigues

This is another of the United States Marine Corps surplus uniforms used by the Republic of Texas, according to Hefter. Both he and Marine Corps artist Colonel Charles Waterhouse show the light blue pants with the dark blue jacket for this time frame, about 1835 to 1842. Smith-Christmas, however, is of the opinion the trousers should be gray with a buff stripe for a sergeant. The inventories suggest that both types of trousers may have been worn by the Texas marines.[79] Nor is it unreasonable to assume that the Texians may have done same alterations in style as might have suited their desire for some difference between their uniforms and those of the U.S. Marines.

The sergeant holds one of the many Patterson colt revolvers issued by the Texas Navy.

Behind is the Hawkins flag.

★ ★ ★

Plate 24 (page 56)

MARINE CORPS

Service Dress – U.S. Surplus 1834

This is still another U.S. surplus marine uniform Hefter said was sold to Texas. According to Smith-Christmas, the Virginia state militia also got some of these surplus uniforms as well as Texas. But he differs again on the color of the trousers. In his color study Hefter shows them to be a lighter shade of green than the coat. Smith-Christmas contends they should be gray. No green pants turn up in the naval clothing inventories, but gray ones do. Thus gray is shown here. The collar and cuff trim, trim on coat tails and the epaulets was buff. The leg stripe for a commissioned officer or non-commissioned officer would also be buff. A green cloth cap is shown, rather than the shako depicted by Hefter, as such caps appear in the naval inventories, whereas no shakos are shown to go with the green uniform in the naval records.[80]

This marine holds a Colt Patterson revolving cylinder carbine, caliber .55, Model 1839-1842. The navy acquired some of these and also some of the hammerless Colt Patterson revolving cylinder rifle, caliber .40, Model 1838-1842. This early use of Colt revolving pistols and long arms demonstrates that Commodore Moore, despite Texas' strained finances, was determined to provide its navy with the latest and most advanced developments in small arms. While the revolving pistols became highly popular over the years, Colt's revolving long arms never caught on. The reason was that a multiple miss-fire on a pistol was not serious; at worst the pistol might be damaged. But with a long arm, because the person firing it was supporting the barrel with one hand forward of the cylinder, a multiple miss-fire could cost him a hand. Some called it "Colt's revolving wheel of misfortune." The Colt carbine shown here, with its Texas star, is in the Winchester Museum, New Haven, Connecticut.[81]

Behind the marine is the ship pennant of the schooner *San Antonio*.[82]

★ ★ ★

There are a number of contemporary references to the uniforms of the Texas Navy confirming that they were worn. Among them is a comment in the press in 1839 when the *Wharton* arrived in New York City that the Texians were in "new uniforms."[83] Other observers were less flattering, describing "officers in tattered dress coats" ... a lieutenant "hocking his epaulet" (a lieutenant wore only one, and on the right shoulder) ... the clerk of the Navy Yard at Galveston "attired in an old tarnished uniform" ... "two midshipmen – without shoes." Moore himself cut an imposing figure with his "imposing sword" and a uniform with a "huge gilt star."[84] The star was not regulation (unless it was a part of the hat insignia), so must have been something he added himself.

There is also a contemporary ink and wash sketch done by Lieutenant Alfred G. Gray, the executive officer of the *Austin*, showing a longboat filled with men of the Texas Navy, all smartly uniformed. Among the uniforms discernible are a captain or high officer in billed cap and blue coat (Plate 17), a midshipman in straw hat and blue jacket (Plate 18), and seamen wearing the summer white uniform (Plate 20) but with dark (red?) vests. Another figure in the front with the officers, in a dark cap and white or gray jacket, might be a marine.[85]

Texas sailors and marines were usually armed to the teeth when making landings on the Mexican coast. Usually each man in a shore party carried a revolving rifle or carbine, a revolver backed up by two single-shot pistols, a sword or cut-and-thrust for officers and a cutlass and knife for seamen. These were sometimes supplemented with "battle axes" and boarding pikes.

Two companies of the newly uniformed First Infantry Regiment, together with militia and volunteers, attacked and defeated the Cherokees in east Texas on July 17. The regulars probably wore the summer uniform described in the text of Plate 9. The Cherokees, who had long been under suspicion of connivance with the Mexicans, now fled to American territory. This was the first of Lamar's Indian battles, which persisted throughout his three-year term, in contrast to Houston's policy of co-existence. But an increase in Indian depredations as Houston's first term drew to a close had turned Texans towards Lamar's militant stance.

Towards the end of 1839 Lamar sent Texas Commissary of Purchases, William Henry Daingerfield, and other agents to the United States to contract for the military uniforms and equipment necessary to fulfill his dream of an empire that would rival the United States.

A contract was made October 15 with William R. Burke of New York for military uniforms totaling $23,389.38. Included were: 225 blue cavalry uniform coats for privates; 20 for sergeants and 25 for corporals; 10 scarlet musicians coats; 260 blue cavalry pants for privates and 20 for sergeants; 235 gray cavalry fatigue jackets for privates and 25 for sergeants and corporals; 560 gray cavalry pantaloons; 280 cotton drill jackets; 560 cotton drill pantaloons; 480 blue infantry coats for privates and 40 for noncommissioned staff and sergeants, plus 40 for corporals; 560 blue infantry pantaloons; 480 gray infantry fatigue jackets for privates, plus 40 for sergeants and 40 for corporals; 1,120 gray infantry fatigue pantaloons; 2,240 cotton shirts; 840 haversacks.[86]

During November Daingerfield signed the following contracts:

- William Cressman of Philadelphia – 560 cartridge boxes with belts for same, having a plate on both belt and box alike; 560 bayonet belts and scabbards with plates; 560 waist belts with plates, etc. and gun slings (like U.S.); 840 leather stocks (like U.S.). The plates were to be of white metal with a raised star of five points.[87]
- George W. Tryon and Sons, Philadelphia, for 560 muskets. This was the Model 1816 flintlock. A few still exist, marked with a star and the name TEXAS.[88]
- John Malseed, also of Philadelphia, for 2,240 laced booties identical to the U.S. Army issue.[89]
- Gilbert Cleland, New York, for 840 forage caps of blue cloth. Of these 280 were to have dragoon buttons (gold) and the balance white metal infantry buttons.[90]
- Magee, Taber (or Faben) and Company, Philadelphia, for 280 saddles with brass stirrups and a like number of bridles, girths, surcingles, cruppers, and holster slips.[91]
- Henry Lutz (in some documents Lutts), Philadelphia, for 840 wooden canteens of one and a half quart capacity, equal to U.S. issue.[92]

By the end of the year, or shortly after, the caps, uniforms, belts, shoes, and canteens had been produced "equal in every respect" to samples or U.S. equipment, according to the Certificates of Inspection done by moonlighting U.S. Army inspectors, and were now ready for shipment to Texas.[93]

An Abstract of Costs incurred by Daingerfield that year shows he also purchased: brown cotton shirts; white woolen blankets; dragoon buttons for coat and vests; infantry coat and vest buttons; dragoon sabres and scabbards; artillery swords; fine guilt mounted leather belts; men's heavy woolen half-hose; plus "saddle, etc., for colonel (Lysander) Wells." This saddle and its accessories, as we shall see in the description of Wells at the Council House Fight the following year was apparently quite an elaborate affair, costing the hefty sum of $193.62.[94]

During the period of the Texas Revolution and republic, the fly on men's trousers was in a state of transition. A French fly (the fly in use today) is mentioned only once in the regulations, for cavalry officers dress uniforms, but was used for most officers uniforms except cavalry officers overalls for stable duty. Nothing is said about which fly enlisted personnel wore. But, like their American counterparts, they wore for the most part a drop fly. The drop fly was a buttoned flap in front (much like the back flap on long underwear). A *short fall* drop fly was nine inches across the front, with three buttons at the top; whereas the *broad fall* drop fly went clear across the front, held up by four or five buttons. What may appear to be a French fly in the illustrations of enlisted uniforms is actually a seam. What might look like pockets on either side are the two sides of the drop fly.

As 1839 drew to a close, Mexican Federalists appeared in Texas seeking military assistance in their intermittent campaigns since 1837 against the Centralists. Lamar refused to commit Texas to an official alliance, but Federalist recruiting took place openly, nonetheless. The Federalists were led by a glib-tongued Laredo lawyer, Antonio Canales, who alternately courted and fought the Texians. His several military adventures into Mexico always ended in disasters. But as a contemporary observed, he "pursued his political ideal with everything but talent."[95] The Federalists wanted a separate nation comprised of the northern Mexican states of Tamaulipas, Nuevo Leon and Coahuila, plus the disputed Nueces Strip which lay between the Nueces and Rio Grande rivers – or so they told the Texians. Actually, they wanted all of Texas as well.

The movement had a certain appeal for Texians, nonetheless. As long as Mexico was entangled in the Federalist struggle, chances for a new invasion of Texas were lessened until the Federalists were dealt with. Plus the Federalists professed to espouse the constitution of 1824, which the Texians had originally risen in revolt to restore.

Canales enlisted two divisions: one of *Tejanos* and Mexicans, the other of "Texian allies." The Anglos, about one hundred strong, were particularly desired as they proved to be the only part of Canales' force that could be counted upon to put up a good fight.

Texas had just organized a new mounted unit, the Gonzales Company, to protect the area west of and below San Antonio. Ruben Ross, a nephew of the filibuster of the same name who took part in the Gutiérrez-Magee Expedition of 1813, was the captain. Ross, charged with the protection of Texas territory, apparently felt that taking sides with the Federalists worked toward that end. After a meeting with Canales, he and his entire command rode off to Mexico. That they considered themselves still in the service of Texas was evidenced by the fact that they rode under a Texas banner and, on their return, Ross and certain others applied for back pay from the Republic of Texas. Disappearing into Mexico with Ross and his command in late September was a certain amount of unspecified "public property." This might have included their uniforms, cavalry equipment and even the flag. The republic sent Colonel Ben Johnson hurrying after them to retrieve the "public property." Not only was his mission a failure, some of those accompanying Johnson defected to Ross.

1840

On January 13, 1840, William G. Cooke, formerly a captain in the uniformed New Orleans Grays during the Texas Revolution, now a Texas Army colonel, succeeded A.S. Thurston as Quartermaster General. Among the public property turned over to Cooke were these uniform items: 1,734 jackets; 2,080 pair of pantaloons; 70 pair of shoes. The jackets and pantaloons, allowing for some being issued by Thurston, were probably the big order from Burke, now paid for and shipped on the schooner *Metron*.[96]

Also headed for Galveston aboard the schooner *Harriet Porter* was a large order of military equipment which included these uniform components: 195 pair half-hose; 840 knapsacks; 2,240 pair laced booties; 560 sets military equipments; 840 leather stocks; 280 pair of spurs; 840 canteens.[97]

Under date of April 16, Cooke directed R.S. Neighbors to Galveston to receive army supplies scheduled to arrive there aboard two schooners. Neighbors was to supplement the two hundred saddles and bridles coming on one of the schooners by purchasing another one hundred "Spanish" saddle trees with "hair girths, surcingles, and heavy stirrups with strong leather stirrup leathers."[98] This ship was without doubt the *Harriet Porter*, whose cargo included military saddles, bridles and wagon harness.

Charles Mason, First Auditor of the republic, recorded in his Settlement of Accounts with Quartermaster General Cooke that Cooke had spent during the first quarter: $5,849.50 on (military) clothing; $2,620 on cavalry equipment; and $695 for cavalry horses.[99]

Supplies of uniform buttons, caps, belts and blankets were also arriving. President Lamar was determined to have the military forces of his fledgling nation as well equipped as those of the United States – *whatever the cost.*

All of the Texas troops engaged in the Council House Fight in San Antonio, March 15, were uniformed. Whether in the fatigue uniform or full dress is unclear. It could have well been in full dress. Author-artist-historian Randy Steffen says that the U.S. Army was well aware of the value of full dress impressing the Indians during negotiations. The Texians may well have followed suit. Tending to support this is the description of Colonel Lysander Wells, head of the Texas cavalry, who was a participant in the fighting. Also, among the companies ordered to supplement the San Antonio garrison during the Council with the Indians was the Travis Guards of Austin who, as documented, had received full dress cavalry uniforms.

The Council House Fight was to have been a peaceful council to make a treaty with the Comanches. A number of the chiefs had been invited. Among the conditions laid down by the Texians was that the Indians bring in all of the White captives they held.

When the Indians arrived at the Council House, however, they had only one. This was Matilda Lockhart, a fifteen-year-old girl who told a heart-rending tale of horrible abuse. Among other tortures during her two-year ordeal as a captive, she had been awakened most mornings by having a hot coal stuck in her nose. The nose was by now almost entirely burned off. The Indians had considered this hilariously funny. Most of the rest of her body was covered with scars from burns as well. Matilda

Lockhart also revealed that the Indians were holding back about fifteen other White captives nearby, intending to gain a better bargaining position by offering them one by one. The appearance of the pathetically disfigured child, her broken spirit, and the obvious deceit by the Indians infuriated the Texians.[100]

When the Texas negotiators confronted the dozen or so chiefs about the other prisoners, the Indians professed to have no other prisoners. Any others were held by other tribes, they claimed. Adding to the annoyance was one chief's smirking question, "How do you like our answer?" At this point a company of uniformed Texas soldiers under Captain George T. Howard was called in. The chiefs were then informed they were hostages until all of the White captives were produced. At this point the interpreter made a hasty, but not unwise, exit. One of the chiefs was right behind him, but was blocked at the door by one of the uniformed soldiers. The chief knifed the soldier and Captain Howard. Weapons appeared from beneath Indian robes and a nasty hand-to-hand fight ensued. Indian warriors outside the Council House fought with another company stationed there under Captain W.D. Redd. When the fight was over Indian casualties were thirty-five dead, including the twelve chiefs. Texas casualties were seven killed, eight wounded.[101]

Nearby was the second highest ranking officer of the army, Colonel Lysander Wells, commander of the cavalry. He was "elegantly uniformed," according to a newspaper report, reinforcing the supposition that the troops engaged were in full dress. Wells was mounted on a "colorfully caparisoned" horse.[102] This must have been the saddle and its accoutrements mentioned in the Abstract of Purchases by Daigerfield. Wells, armed with a Patterson Colt revolver, managed to get the best of an Indian who jumped up behind him on his horse, despite the malfunction of his pistol.[103]

The lure of adventure with the Federalists was a siren's song for the restless soldiers of Texas. Early in the year President Lamar had found it necessary to issue a proclamation to the army forbidding enlistments with the Federalists. It did little good. In addition to offering the same pay as the Texas Army (which paid the same as the American army when it could), Canales dangled the enticement of a share of the spoils. Shortly after the Council House Fight the commander of the San Antonio garrison, former acting Secretary of War William S. Fisher, without resigning from the army, raised two hundred men for Canales. Most were from the San Antonio garrison.

Since these troops had all been fully uniformed at the Council House Fight, did they ride off with the Federalists in their Texas Army uniforms? Very likely. What other clothing did they have? Reading and re-reading of the military experience in Texas, it is always amazing how poorly clothed and armed were the men who responded to any call to arms. There was always a want of clothing, even shoes, and many came without even firearms. There seemed to be only one indispensable item, according to the critical traveler Colonel Edward Stiff, "the savage Bowie knife, which as if by common consent, was a necessary appendage to all."[104]

In June, Major Neighbors was authorized by Quartermaster General Cooke to outfit a new military unit, the Border Guards, raised at Galveston, with "quartermaster stores and ordnance."[105] It would not be unreasonable to assume these supplies included uniforms, since the Burke order and others had arrived in Texas and *by now there were more uniforms than soldiers*.

The next month Charles Mason did an audit of Cooke's accounts for the quarter ending June 30, showing he had disbursed monies for: (military) clothing $1,311.22; cavalry equipment $100.90; cavalry horses $35,100.[106]

The Comanches began a series of raids as revenge for the massacre of their people at the Council House in San Antonio. These culminated in an assault on Victoria and the burning of Linnville. The hundreds of warriors that took part made a fateful tactical error, in the opinion of Major General Thomas S. Bishop, a serious scholar on Texas military history. They had raised a hornets nest of mobilized Whites in their wake, and now chose to exit by the same route.[107] Regulars led by General Felix Huston, no doubt still nattily attired *a la militaire*, assisted by militia, rangers and volunteers, gave them a sound defeat at Plum Creek, August 12. It was the greatest Indian battle in Texas history, certainly the most colorful. In contrast to the drably uniformed Texas regulars and the plainly dressed militia (some of whom may have been uniformed), volunteers and rangers, the sight of the Comanches was unforgettable according to a Plum Creek participant:

> "It was a strange spectacle never to be forgotten, the wild, fantastic band as they stood in battle array, or swept around us with the strategy of Indian warfare ... *both horses and riders were decorated most profusedly, with all the beauty and horror of their wild tastes combined*. Red ribbons streamed out of their horses' tails as they swept around us, riding fast, carrying all manner of stolen goods upon their heads and bodies.
>
> "There was a huge warrior, who wore a stove-pipe hat, and another with a fine pigeon-tailed coat, buttoned up behind. They seemed to have a talent for finding and blending the strangest, most unheard of ornaments. Some wore on their heads immense buck and buffalo horns. One headdress ... consisted of a large white crane with red eyes ..."[108]

Shortly after, the Comanches were dealt another serious defeat when their main village on the upper Colorado was destroyed by pursuing Texians. The Comanches thereafter stayed away from the larger settlements, harassing the frontier only, mainly to the west.

By now the disillusioned Ruben Ross and the survivors of his company had dragged back from Mexico, wiser and poorer from their experience with Canales' Federalists. Not only had Canales welched on payment, but their applications for back pay from Texas were denied. Canales found new suckers, however. Joining the two hundred Anglos recruited by Fisher were about 110 more under Samuel W. Jordan, who was given the rank of colonel. About two hundred *Tejanos* were assembled by Juan Seguín, former lieutenant colonel of the Texas Army, senator and mayor of San Antonio. There was also an undetermined number of Mexican *rancheros* plus some eighty-seven Indians. It was a rough, polyglot lot. The same pay and booty, as before, were the lure, albeit the former had never been fully paid. Again, the adventurers rode into Mexico, their general,

Canales, always managing to lag behind and somehow miss the fighting which, again, was done mainly by the Texians.

The demise of Canales' Republic of the Rio Grande came as a surprise to the "odious Texans," as the hitherto indispensable allies were now called by Canales' brother-in-law Juan Molano, who headed the Mexican unit. After the unsuspecting Texians had been lured deep into the interior of Mexico the Federalists made a pact with the Centralists, joining them. One of the conditions for the union was that the Federalists "separate from, and abandon to their fate, the adventurous strangers at present among them."[109] When the "adventurous strangers" finally caught on, they managed to escape the trap they had been led into and miraculously fought their way back to Texas.

At the border, the returning Texians found Seguín and his *Tejanos* safely on the Texas side. They had chosen not to get involved in the final debacle, whether through prudence, luck or pre-knowledge of the secret deal. Seguín then left for Monterrey where he dallied for some months trying, he claimed, to collect the pay due himself and his men from the Centralists. As part of the pact the Centralists were to pay off all Federalists for the time they served the Federalist cause. They were also to retain their military ranks. But the Centralists double-crossed both Texians and *Tejanos*. Seguín was allowed to return unmolested, though empty-handed. But the few gullible Texians who re-crossed the Rio Grande into Mexico to collect their pay under the agreement were arrested and imprisoned. The duplicity of Canales and suspicion about the role played by Seguín further aggravated the relationship between Texians, *Tejanos* and Mexicans. The entire affair, many felt bitterly, was another example of Mexican double-dealing. It was a lesson, one contemporary commented, for those who were "rash enough to put faith in a misguided Mexican."[110]

Perhaps the loss of the "public property" that had been taken into Mexico by the Texas soldiers who took "French leave" to join the Federalists, plus the Comanche fights, not to mention the big orders for uniforms, accounted at least in part for the heavy expenditures from October 1, 1839 to September 30 the following year. The Exhibit of Monies received and disbursed by Quartermaster General Cooke shows: (military) clothing, knapsacks, haversacks, etc. $69,470.54; cavalry equipment (exclusive of horses) $30,382.[111]

Near the end of the year William L. Cazneau wrote to Daingerfield explaining that uniforms and equipment requisitioned by Cooke for troops stationed at posts on the Red River and Trinity River were in stock in the quartermaster stores at Austin, but there were no means of transporting the items. The requisition called for: 168 woolen overalls (synonymous in those times with trousers); 168 woolen jackets; 504 cotton overalls; 168 cotton jackets; 672 cotton shirts; 504 pair of shoes; 168 forage caps; 504 pair of woolen socks; 12 gross infantry buttons, large size; 6 gross infantry buttons, small size; 168 blankets; 48 camp kettles; 24 mess pans.[112]

As the year progressed Daingerfield contracted with others:

> Andrew Fenton for knapsacks and tents; Thomas R. Fisher for halfhose; Alfred Hursts and Scovil for buttons; Samuel Lloyd for camp equipment; N.P. Ames for sabres, swords and scabbards; Henry H. Williams for nails, hammers, spades and axes; Wilson Childs and Company for wagons; Mathias Seddinger (in some documents Matthias or Sedding) for spurs.[113]

The navy in the meantime was uniforming and equipping its men separately. It was not until later that the army and navy departments were combined into one War and Marine Department.

In January the public stores at the Navy Yard had then shown: battle axes 62; marine belts 51; green cloth caps 8; soldiers dress caps 43; soldiers fatigue caps 31; marine dress coats 45; pea coats 219; flannel drawers 24 pair; black silk handkerchiefs 15; blue cloth jackets 254; (illegible) cassinette jackets 14; fatigue jackets 103; monkey jackets 50; new jackets 68; cadet cloth pantaloons 68; while cotton shirts 57; flannel shirts 400; shoes 60; soldiers stocks 121; blue cloth trousers 171 pair; duck trousers 1. Also on hand, no doubt to keep morale afloat, was more than 386 gallons of whiskey and rum.[114]

By mid-summer the inventory was less spare: blue cloth caps 89; marine green cloth caps 10; soldiers dress caps 125; soldiers fatigue caps 125; 8 bales of clothing; marine dress coats 127; soldiers overcoats 100; pea coats 381; 2 boxes of cutlasses; flannel drawers 404 pair; duck frocks 1,031; black silk handkerchiefs 28; woolen hose 120 pair; blue cloth jackets 470; cassinette jackets 120; fatigue jackets 128; 2 boxes of muskets; 26 boxes of old muskets; fine white pantaloons 120 pair; cadet cloth pantaloons 130 pair; cassinette pantaloons 148 pair; soldiers white cotton pantaloons 143 pair; nankeen pantaloons 36 pair; pompoms 125; red flannel shirts 19; white flannel shirts 90; white cotton shirts 161; shoes, brogans 120 pair; low quartered shoes and pumps 63 pair; soldiers stocks 200; blue cloth trousers 503 pair; duck trousers 750 pair; blue drilling trousers 170.[115]

A printed form for the quarter from June 12 to September 30 of Slops and Small Stores held by the Naval Storekeeper showed: pea jackets 2; fine pea coats (probably for officers) 2; blue cloth jackets 26; blue cloth trousers 22; flannel shirts 40; guernsey frocks, none; flannel drawers 40; woolen socks 21 pair; blue cloth caps, none; duck frocks 34; duck trousers 43; shoes 16 pair; naval coat buttons 44 dozen; naval vest buttons 60 dozen; sheath knives 1; hands of grass 91.[116] Slops were articles of clothing or bedding issued to sailors from a ship's stores.

There were additional stores of uniforms aboard the vessels, as evidenced by pursers records of the various ships. A Return of Slops report for the quarter ending September 30 shows that the purser aboard the *San Antonio* had: pea jackets 22; fine pea coats 1; blue cloth jackets 30; blue cloth trousers 26; flannel shirts 47; super overcoats (perhaps like the ankle-length one shown in the Johns sketch?) 2; flannel drawers 41; woolen socks 20 pair; blue cloth caps 1; duck trousers 33; pieces of ribbon (possibly for hat bands?), none; blue dungaree trousers 2; fine white pants (probably for officers) 1; nankeen pants 10; white cotton counterpanes, none; colored counterpanes, none; black silk handkerchiefs 77; shoes 11 pair; naval coat buttons 47 4/12 dozen; naval vest buttons 15 8/12 dozen; sheath knives 37; hands of grass 600.[117]

Aboard the flagship *Austin*, personally commanded by Moore, a year-end Return of Slops and Small Stores showed:

pea jackets 100; fine pea coats, none; blue cloth jackets 200; blue cloth trousers 200; flannel shirts 200; guernsey frocks, none; flannel drawers 200; woolen socks 216 pair; blue cloth caps 20; shoes 121 pair; brogans 25 pair; pumps 32 pair; naval coat buttons 168 dozen; horn or bone buttons, none; sheath knives 106; duck frocks 335; duck trousers 200; blue dungaree trousers, none; fine white pants 40; nankeen pants, none; white cotton shirts 2; colored shirts, none; black silk handkerchiefs 240; hands of grass 600.[118]

By now the army and navy of the Republic of Texas were smartly uniformed. They were not the rustics in home-spun costumes depicted by movie makers and even the majority of historians.

1841

On March 2, 1841, a special order from Secretary of War Branch T. Archer advised Colonel William G. Cooke that his command, the First Infantry Regiment, was to be disbanded by act of Congress.[119] This was part of an economic cutback.

But President Lamar wasn't to be deterred from his dream of empire. Without Congressional approval he began implementation of his most ambitious project, the Santa Fe Expedition. Its announced purpose was to establish trade with the territory of New Mexico, claimed by Texas but still under Mexican control. A secondary purpose, and not a well-kept secret, was – if conditions proved receptive – to annex the territory.

The bulk of the Santa Fe Pioneers, as they were called, was a military force of some 270 mounted soldiers plus General Hugh McLeod and his staff. They escorted three commissioners sent by Lamar to negotiate trade, annexation or, ideally, both. These were José Antonio Navarro, William G. Cooke (who later married Navarro's niece Angela) and R.F. Brenham. The balance of the more than three hundred participants were merchants with wagons of merchandise to open Texas' Santa Fe trade. Accompanying the Texians was an American journalist, George Wilkins Kendall, editor of the *New Orleans Picayune*. Kendall left a first-hand account of the ill-fated expedition.

All of the military was smartly uniformed. Even two of the three commissioners were in uniform as well, Cooke and Brenham. Arms, horses and horse equipment were all government issue. The General Abstract of Quartermaster Stores drawn by officers of the Santa Fe Expedition, signed by William L. Cazneau on July 1, included: blue cavalry coats 233; blue infantry coats 1; musicians (red) coats 3; forage caps 215; canteens 293; haversacks 234; gray cavalry jackets 148; white cavalry jackets 55; knapsacks 177; blue cavalry pants 182; gray cavalry pants 522; white cavalry pants 251; shirts 369; shoes 740; socks 712; stocks 112; saddles 184; pairs of spurs 118; bridles 181; Colt rifle 1; violin strings 12 (apparently at least one of the musicians was playing something other than a bugle). In addition, there are numerous individual requisitions in the archives for various bits of uniform and equipment.[120]

The military escort consisted of six companies. Each was followed by a wagon containing its supplies. Some were cavalry, some dragoons (mounted infantry) and one was artillery with a single mule-drawn six-pounder cannon.

The expedition assembled on Brushy Creek near Austin. There they were reviewed by President Lamar and other officials.

In June they set off for Santa Fe. The journey was long, and though they were well mounted and provisioned initially, men and animals soon began to wear down as the country began to grow more and more inhospitable. They lost their guide, then their way. At last they reached New Mexico but in deplorable condition. Waiting for them with his militia was New Mexico Governor Manuel Armijo. He captured all of the exhausted Texians by trickery without firing a shot. They were marched to imprisonment in Mexico. Many died along the way from brutal treatment, thirst and exhaustion.

In his book, Kendall mentions the uniforms frequently. When the artillery officer, Lieutenant William F. Lewis, tried to pass off the expedition as American merchants, Armijo wasn't fooled. Seizing Lewis by his coat with its Texas buttons, Armijo demanded, "What does this mean? You need not think to deceive me: no merchant from the United States ever travels with a Texas military jacket."[121]

Kendall also gives a good description of the uniform of Second Lieutenant C.C. Hornsby of the cavalry. Hornsby, he said, "was the best dressed man in the party of prisoners. His well-wadded and full buttoned Texas dragoon jacket was new, or nearly new, his cap and military trousers had seen but little service, while his blanket was of that fiery red which could not be found in a country where gaudy and glaring colors are so sought after and admired." Hornsby's dragoon jacket and cavalry blanket were so coveted by a New Mexican civilian who observed him as the Texian prisoners were marched toward Mexico that, alas for Hornsby, he was forced into an involuntary exchange for the ill-fitting jacket and course blanket of the admirer of his finery.[122]

In other references to the uniforms, Kendall mentions how the soldiers were forced to barter off their garments to sustain themselves. The first item to go had been the buttons. Apparently buttons were in short supply then in New Mexico and much sought after. They were durable brass and white metal buttons: one wonders if any still exist, now possessed by persons in humble circumstances not knowing their historic value. As their arduous march continued one might see "a pair of ragged Mexican-made trousers ... a faded Texan dragoon jacket ... a complete outfit ... then a man in a dragoon cap worn jauntily on his head ... part of a shirt and occasional fragments of what had once been military pantaloons." By the time they passed beyond El Paso nine out of ten had traded off their uniforms.[123]

The expedition was well armed, according to Kendall. This is evidenced by the records. Among the long arms were a number of the Tryon flintlock muskets. These were probably those originally issued to the now disbanded First Infantry Regiment. This musket was intended to be the principal arm of the Texas Army. It was probably the arm of the mounted infantry on the Santa Fe Expedition, many of whom were recently discharged from the First Infantry regiment.[124]

Among long arms ordered and received for the cavalry were 250 Jenks patented carbines. These arrived on the schooner *Neptune* at Galveston, April 17, 1841, thus in time to accompany the Santa Fe Expedition.[125] They may have been the carbines mentioned by Kendall. There is also evidence that among the long arms on the expedition was at least one Colt sixteen-shot revolving rifle. The Republic had also bought 200 to 250 of these. A twin-barrel shotgun is also mentioned by Kendall. Such shotguns were also government regulation.

Hand guns carried to Santa Fe included a brace of heavy horsemen's pistols for each man. Some, especially officers, were armed with the latest in pistolry, Colt's five-shot Patterson revolver. The Pattersons had come from the navy. According to Commodore Moore, "In the year 1841 I furnished the Santa Fe Expedition with forty pistols belonging to the schooner *San Jacinto*, which vessel was lost in the Arcas Islands in November 1840." Records of the *San Jacinto* show that at least half of those pistols were Colt Patterson revolvers.[126]

Most of the Pattersons were destroyed by the Texians upon capture rather than let such a valuable and formidable weapon fall into Mexican hands. But not all. After his capture, Kendall mentions seeing in the waist sash of one captor, "two of Colt's revolving pistols taken from one of my friends."[127] These must have been the pair formerly possessed by Major George T. Howard, whom Kendall had said earlier had a pair. This was the same Howard who had been knifed by an Indian in the Council House Fight.

All of the military equipment of the Santa Fe Expedition, uniforms, rifles, shotguns, muskets, saddles, bridles and even the six-pounder cannon and its new mule harness was lost. This was a heavy blow to a government already financially strapped, and devastating to Lamar's vision of a militaristic empire.

Navy store at the beginning of the year showed: battle axes 62; marine belts 51; green cloth caps 8; soldiers dress caps 43; soldiers fatigue caps 31; marine dress coats 45; pea coats 219; soldiers overcoats 88; black silk handkerchiefs 15; blue cloth jackets 254; casinet jackets 14; fatigue jackets 103; monkey jackets 50; new muskets 68; pantaloons of cadet cloth 12; casinet pantaloons 33; white cotton pantaloons 68; pompoms 43; flannel shirts 400; white cotton shirts 59; duck trousers 1; blue cloth trousers 171; shoes 60 pair. Undated, but from roughly the same time frame is a Statement of Receipts and Issues of John G. Tod "late naval storekeeper" showing a balance on hand of: marine coat buttons 40 1/2 gross; marine vest buttons 37 gross; navy coat buttons 43 gross; navy vest buttons 81 gross; sailors buttons 12 gross; green cloth caps 31; soldiers dress caps 43; pea coats 157; pea coats damaged 1; soldiers overcoats 88; red flannel (illegible) 4; duck frocks damaged 1; duck frocks 79; black silk handkerchiefs 15; black silk scarves 7; blue cloth jackets 198; casinet jackets 93; white fatigue jackets 93; monkey jackets 50; muskets 68; casinet pants 22; white cotton pants 36; flannel shirts 3; soldiers stocks 121; duck pants 1 pair.[128] Some of the items on the Tod list, but by no means all, appear to be carried over from the January list of naval stores.

Records of the ship pursers and of the public stores at the Texas Stores and Navy Yard at Galveston show a continuous issuance of uniforms, buttons and other slops throughout the year.

Added to these inventories were purchases by Commodore Moore during the quarter ending June 30; marine casinet trousers 13; white marine jackets 93; marine white pants 50; watch coats 88; dress coats 45; dress caps 43; foraging caps 31; officers caps 8; muskets 68; marine belts 51; blue cloth jackets 186; pea coats 41; duck frocks 43; monkey jackets 50; red flannel shirts 4; coat buttons 43 gross; vest buttons 63 gross; marine coat buttons 40 gross; marine vest buttons 37 gross; sailors buttons 12 gross; navy vest buttons 37 gross.[129] Some of these purchases may have been out of Moore's own pocket, or on his pledged personal credit.

Lamar thought of the navy as a possible source of revenue rather than a necessary financial drain. Opportunity presented itself when Colonel Martín Peraza turned up in Austin to negotiate a treaty between Texas and Yucatan, which was now in revolt against the central government of Mexico. In September an agreement was signed whereby Yucatan would rent the Texas Navy for $8,000 per month. This was just in time, for the Congress had also cut naval appropriations. Also in September, Sam Houston, a confirmed navy hater, was elected to his second term as president.

On December 13, Lamar left office and Houston took over. Predictably Houston, as one of his first acts, rescinded Lamar's authorization for the navy to sail for Yucatan. He was too late. Commodore Moore, wise to Houston's ways, had cleared port with his ships, officers and men smartly attired in the uniforms prescribed by the regulations of 1839. They didn't come back until May of the next year. While Houston fumed, they made naval history.

1842

In the first months of 1842, Houston reversed the military policies of Lamar. In addition to his thwarted effort to curtail the navy, he sought peace with the Indians. They were glad to accede after the drubbing they had received from Lamar's military. Indian depredations decreased, but whether the credit belongs to Lamar or Houston is debatable. Though not in concert, they had successfully worked the "bad cop-good cop" routine on the Redmen.

On February 28 and March 1, auctions were held of army quartermaster stores by order of Houston's Secretary of War. Items offered for sale were: cavalry cotton pantaloons 102 pair; infantry cotton pantaloons 29 pair; infantry and cavalry cotton jackets 70; haversacks 13; blue infantry coats 106; musicians coats 2; gray cavalry pantaloons 43 pair; gray infantry pantaloons 50 pair; cavalry and infantry jackets 14; common woolen jackets 3; blue cavalry pantaloons 12 pair; woolen socks 60 pair; leather stocks 72; infantry buttons 9 gross; brogans 89 pair; knapsacks 200; tin mess pans 67; tin mess pans (old) 70; box with old iron and bugle 1; drums (old) 2; blue infantry pants 25 pair; barrel of leather straps 1; blankets (old) 4; saddles (old) 74; one lot of 72 saddle pads.[130] It would appear that only a small part of these items were old. So it is unlikely that this represents only a disposal of worn out or unserviceable items. More likely the auctions represented the wish of Congress and President Houston to downscale the regular armed forces and depend for defense once more primarily on militia. But this is unclear.

Despite Houston's hope for peace, conflict increased. on March 3 a small party of Mexican regulars seized Goliad. Two days later San Antonio was taken by surprise and occupied by a force of between 500 and 700 Mexican soldiers under General Rafael Vásquez. Refugio was also occupied the same day. Houston called out the militia. But by the time 3,000 Texians turned up at San Antonio to confront the invaders, the Mexicans had departed. With them went Juan Seguín and a number of *Tejanos*, now disenchanted with Texian dominance.

Among the militia that rallied, at least two units were in uniform. Both, the Galveston Coast Guards and the Galveston Fusiliers, were from Galveston, then the largest city in Texas. We have a clear description of what the Coast Guards looked like from an English visitor, William Bollaert, who accompanied them as they searched for Mexican invaders.[131]

★ ★ ★

Plate 25 (page 57)

GALVESTON COAST GUARDS (the Sea Fencibles)

News of the Vásquez invasion in 1842 produced "intense excitement" in Galveston. Public meetings "attended to suffocation" mustered militia units, most prominent of which was the Galveston Coast Guards.[132]

The Coast Guards, one hundred strong, were nattily turned out "with an eye to uniformity" in straw hats, red woolen shirts and white pants. Their arms were muskets and pistols, pikes and swords "but there was no scarcity of hatchets, tomahawks, and Bowie knives for close quarters if necessary."[133]

On March 14, the Coast Guards, together with the fifty Galveston Fusiliers, set forth in the steamer *Lafitte*, accompanied by two small craft of the Texas Navy. They searched the coast for a rumored Mexican invasion fleet. A third group, the Galveston Guard, was left behind to defend the women and children.

Among the donations received by the volunteers from local citizens were a plentiful stock of ammunition, provisions and "small stores." These included Honey Dew tobacco, "segars" and "sundry other comforts." Among the sundry other comforts was a generous supply of liquor. The Coast Guards by unanimous resolution urged that the liquor be disposed of to buy more essential equipment. But somehow the liquor found its way aboard.[134]

As they cruised the coast, alert for the non-existent Mexican invasion fleet, the volunteers fought boredom with song and the sundry other comforts. While champagne corks went "pop-pop-pop" and "bright claret" and whiskey punch flowed, the Englishman composed the *Red Rover Song*. It extolled the patriotism and prowess of the Coast Guards and warned of the doom awaiting any Mexicans they should meet. His shipmates, highly pleased, christened him "Blowhard" and promoted him from "waister" (a semi-useless helper) to sergeant.[135]

Some three days after the Sea Fensibles (as they were also known) had "quaffed with ecstasy the last of the glorious liquid," they managed to lay their hands on a lone Mexican. Despite his protest that he was only a herdsman, the otherwise empty-handed warriors pronounced him to be a spy and prepared to take him back to Galveston as the single tangible trophy of their campaign. The "wily Mexican" proved to have too much "smartness," however. On the pretext of rounding up some beef for the return trip, he borrowed a fine riding mule, saddle and a lasso and disappeared over the horizon never to return. This left the trusting Texians "in no very good humor."[136]

★ ★ ★

As to what the uniforms of the Galveston Fusiliers looked like, Bollaert gives us little information to go on except that they were "habited in modest uniforms" and armed with patent rifles with bayonets. These patent rifles were described as "a rather queer looking breech-loading" rifle of superior type. They were probably some of the 250 Jenks bought the year before by the republic, described in one document as carbines and in another as rifles. The fusiliers also drilled with boarding pikes.[137]

The Galveston Hussars Cavalry Company was also mobilized. They proceeded overland, joining a group of about one hundred men led by former President Lamar and General Albert Sidney Johnston. This group was also armed with Jenks. They must have been a special elite to have been armed with the latest in long arms, for the Jenks was a far superior weapon to most used by the republic. Whether the Galveston Hussars or the unit under Lamar and Johnston were uniformed cannot be determined.[138]

Two highly interesting documents, until now unnoticed by others apparently, repose in the Texas State Archives, disclosing that *the militia also had government issued uniforms*.

One is undated but, because of the items listed in it also appear in the other one (which has a date of April 30, 1842) must be presumed to be in more or less the same time frame. It begins: "Received of William L. Cazneau, Quartermaster and Acting Commissary General of the Militia, the following quartermaster and commissary stores for which I am accountable: 171 pair cotton drill cavalry pantaloons; 61 cotton drill cavalry jackets; 18 cotton drill infantry jackets; 33 pair cotton drill cavalry pantaloons (these must have been of a different color than the 171 pair); 26 haversacks; 33 blue dragoon coats; 132 blue infantry coats; 4 red musicians coats; 43 pair gray cavalry pantaloons; 52 pair gray infantry pantaloons; 15 gray infantry jackets; 13 gray cavalry jackets; 13 pair blue cavalry pantaloons; 60 pair woolen socks; 324 leather stocks; 19 gross dragoon coat buttons; 53 gross infantry coat buttons; 15 gross vest coat buttons; 9 gross plated infantry staff buttons; 97 pair coarse brogans; 248 knapsacks and straps; 1,186 new tin cups; 26 pair old infantry blue pantaloons; 1 barrel leather straps; 4 old blankets; 10 stirrup irons; 2 new cavalry saddles; 72 saddle pads; 1 box old iron and bugle; 2 old drums; 2 branding irons; 26 pair old infantry blue pantaloons; 1 new American saddle, damaged and without stirrups; 38 old American saddles, and leathers."[139] After solemnly agreeing to be responsible for the uniforms and equipment, the recipient didn't sign his name! The other document, however, suggests he was Jacob Snively.

The April 30 document, which may be before or after the undated one (it is unclear) shows a transfer of commissary stores from Cazneau, as Quartermaster of *Militia*, to Jacob Snively, Acting Quartermaster and Commissary of the *Army*. These items are almost exclusively uniforms: cotton drill cavalry pants 171; cotton drill infantry jackets 61; cotton drill infantry jackets 18 (again, perhaps a different color than the 61 jackets); haversacks 26; blue cavalry coats 33; blue infantry coats 132;

red musicians coats 11; gray cavalry pants 43; gray infantry pants 52; gray infantry and cavalry jackets 28; cavalry white jackets 3; blue cavalry pants 12; woolen socks 60 pair; leather stocks 324; dragoon coat buttons 19 gross; infantry coat buttons 19 gross; Small vest buttons 15 gross; plated staff (buttons) 9 gross; brogans 97 pair; knapsacks 248; old infantry blue pants 26; stirrup irons 10; new cavalry saddles 2; old saddles 39; saddle pads 72; wood canteens 80; cotton drill pants 33; tin cups 486; old drums 2; old iron and bugle 1.[140]

The importance of these two inventories is that they document for the first and only time that a certain amount of uniforms had been set aside for issuance to the militia. Until their discovery by the author, it had been assumed that (except for special units like the Milam Guards or Travis Guards that were presidential escorts) the militia either provided their own uniforms (like the Galveston Coast Guards) or wore their own civilian clothing.

Exactly what else these transfers represented can only be conjectured. It might mean that no more uniforms were to be issued to the militia. Or it might simply have been that the army stores were critically low at that time. For Snively reported in a separate list that the army's supply of quartermaster and commissary stores on April 30 to be only: cavalry coats 33; white cavalry pants 35; shoes 6 pair; musicians coats 2; haversacks 13; knapsacks 48; tin cups 384; mess pans 54; canteens 56.[141] A third explanation, and perhaps the most likely, is that by now the separate departments of War and Marine had been combined into one Department of War and Navy. All this was probably part of the government's budget cutbacks.[142]

In anticipation of further intrusions by Mexico, a volunteer unit was assembled under General James Davis at Camp Lipantitlan, near the old Mexican earthen fort of Lipantitlan close to San Patricio. The unit was referred to as an "army" in correspondence from headquarters at Austin, but Davis, in his correspondence calls it a "battalion." Considering the size – not quite two hundred men – Davis' description is the more realistic. They were mostly new volunteers from the United States. Houston had required that any newly recruited volunteers from the states must bring their own equipment and provisions. According to W.J. Mills, Inspecting and Mustering Officer, the supplies in possession of the new arrivals, who numbered 165 officers and men, were: swords 18; pistols 51; rifles 48; shotguns 15; cartridge boxes 68; powder flasks 78; knapsacks 84; pounds of powder 579; pounds of lead 783; flints 1,230; haversacks 122; canteens 164.[143] There is no mention of uniforms by Mills but the quartermaster correspondence related to the "Volunteer army" under Davis shows that Snively authorized them to be supplied out of the stores on hand at Galveston.[144]

In June this force was attacked by about seven hundred or so Mexicans under General Antonio Canales, the former kingpin of the Republic of the Rio Grande. About five hundred of Canales' soldiers were mounted and he had some artillery. The clash was brief. There was little loss on the Texian side except for the flag of the Invincibles, left behind at an abandoned camp. This flag – with the inscription *Galveston Invincibles - Our Independence* – is now in the Museum of History in Mexico City, according to Hefter.[145] While the outcome was indecisive, Canales withdrew to Mexico.

A more serious raid, in which San Antonio was again captured by surprise came on September 11. The fifty Texian regulars stationed there made a hasty retreat when faced with 1,400 Mexican regulars under General Adrian Woll, a foreigner in the service of Mexico, described as a Frenchman or Swiss, who had served against Texas during the Texas Revolution. Woll had cavalry, infantry and artillery plus about two hundred mounted *Tejanos* who had been sulking in Mexico. The *Tejanos* were under the command of former Texas Lieutenant Colonel Juan Seguín. Tagging along also was Vicente Cordova, leader of the Cordova Rebellion of 1838-1839, which had stirred up a few Indians and disgruntled Mexicans against Texian rule.[146] This incursion was to have serious consequences.

During Woll's occupation, Seguín was particularly obnoxious. His men killed three civilians bathing in a creek near San Antonio, including Dr. Lancelot Smither who had served on the city council while Seguín was mayor. They also were accused of confiscating property of Texians and *Tejanos* indiscriminately.

At Salado Creek Seguín's mounted unit led an attack on a detachment of Texians under Matthew ("Old Paint") Caldwell which was headed for a rendezvous with other defenders. Caldwell's men, with an advantageous position, inflicted about one hundred casualties on Seguín and forced him to retreat. Not so lucky was another group under Nicholas Dawson. Only two miles away, they were devastated by fire from Mexican artillery and wiped out with a loss of fifty-three killed and ten captured. Only two escaped. Seguín shared the onus of the Dawson massacre, as it was known. Certain Colt revolving arms captured from the Dawson men were sent to Santa Anna by Woll as a special gift.

Again the assembled Texians were too late. Woll and his minion Seguín, after only a few days in the city, retreated back into Mexico. This time, however, the Texians were in no mood to return to their homes. Houston bowed to popular indignation and called for a declaration of war. He ordered General Alexander Sommervell to march the army to the border. But it was all for show. When Congress gave him the declaration of war, he vetoed it.

In the meantime, after a lot of foot-dragging, Sommervell finally reached the Rio Grande. From Laredo he proceeded upstream to Guerrero. There, feeling public outrage had cooled sufficiently, Houston ordered their return. Most went back. But six companies – about 260 men – still steaming to carry the fight into Mexico, elected William S. Fisher their commander and marched on Mier. This was the same Fisher who had formerly been Acting Secretary of War and who had been one of the Texian mercenaries under Canales in the last Federalist debacle in Mexico.

Attacking Mier on Christmas day, The Texians got more than they bargained for: 3,000 just arrived Mexican regulars. After giving a good account of themselves for most of the next day they were nevertheless forced to surrender when their ammunition and provisions ran low. Like the Santa Fe prisoners, they were marched overland to prison deep into Mexico, in fact the same prison, the Castle of Perote.

Were any of the Mier men uniformed? Most were not, but there is some evidence that a few were. Part of Houston's strategy for chilling the ardor for invasion of Mexico was to be niggardly with supplies. Clothing and rations were not forthcoming. The men had to make do as best they could. While they were assembled near San Antonio, Thomas Jefferson Green, later one of the leaders of the Mier faction, observed, "In the shortest possible time they transformed the covering of many an unwary buck to their own legs ... the leather breeches-making went on cheerily ... it looked like a preparation for a tournament in which every man was required to be clad in deer-hides."[147] Many of these costumes, incidentally, were on the pattern of the hunting costume that was the cloth fatigue uniform of the Alabama Red Rovers during the Texas Revolution, but undyed. The sketches of one of the Mier men, Charles McLaughlin, that illustrate Green's book about the Mier fiasco show a great many outfitted as Green described. But McLaughlins' sketches also suggest a few might have been uniformed. In his drawing of the prisoners drawing the beans that meant death for every tenth man as punishment for an escape that succeeded for a few days, can be seen several of the Texians in what could be Texas fatigue uniforms, though with felt hats instead of caps. While caps were prescribed headwear on the Santa Fe Expedition as well, the sketches illustrating Kendall's book also show some of the pioneers wearing felt hats instead of caps. More convincing, however, and leaving little doubt, is the sketch of the breakout during the escape. Far to the right is a Texian who may be in uniform carrying off an ammunition box, but in the right center a Texian, ramming down a load into a captured Mexican musket, *is wearing the complete Texas Army fatigue uniform, including cap* (see Plate 9).[148]

On the Mier raid the Texians had a blood-red silk flag with a white star, which Green carried at times in his hat. This was not an official Texas flag, however. It was not captured at Mier as it had been left behind with the camp guard on the Texas side of the river, who escaped. The Mier flag reappeared briefly during the American Civil War when it was flown over the Coryell Country Court House as a secessionist banner. It finally disintegrated from wind damage, a deplorable loss of a relic that should have been preserved.[149]

1843

The dual disasters of Santa Fe and Mier had Texians seething for revenge against the Mexicans. Early in 1843 Jacob Snively proposed a third expedition. This one would journey northward into the remote area where Texas, New Mexico and the United States came together. There it would waylay some of the caravans of Mexican traders plying the trail from Santa Fe to Saint Louis, Missouri. The spoils would be shared jointly by Snively's adventurers and the Republic of Texas. In February the government approved his proposal.

In April Snively set off with his privately recruited Battalion of Invincibles, about 177 men. Many were former Texas Army soldiers. On reaching the remote corner of Texas over which some of the Santa Fe trade passed, they had a clash with about one hundred Mexican soldiers. The Mexicans lost seventeen killed and eighty-two captured, the Texians none. But no caravans were immediately forthcoming. Some expedition members began to get discouraged. Dissension and desertions weakened their ranks. The final blow was the arrival of three hundred United States dragoons with a cannon, who had been sent to protect the caravans from the Texians. They surrounded Snively's group and disarmed them. By July the adventurers had returned in failure. After protests from the government of Texas it was ascertained that Snively's force had indeed been on Texas territory, not American as the leader of the dragoons had claimed. Compensation was paid to Texas by the United States and the arms that the expedition members had surrendered were returned.

Were there uniforms on the Snively Expedition? Some of the former Texas Army soldiers may have retained their old uniforms, but the government's acceptance letter from M.C. Hamiltion, Acting Secretary of War and Marine, suggests none were issued for the campaign. It included these conditions:

> "The expedition will be strictly partisan, the troops composing the corps to mount, equip and provision themselves at their own expense ... the government to bear no expense whatsoever."[150]

In the records for the third quarter of the year is a receipt showing that Major J.W. Tinsley received from M.J. Warfield for use of the cavalry: 30 hunting shirts; 10 R(oundabout) jackets; 18 shirts; 118 pair of pants; 6 coats; 15 vests; 104 pair of shoes; 56 hunting shirts (which must have been in some way different from the other thirty hunting shirts?); 43 caps.[151] Notice the official preference for caps over felt hats. The contemporary sketches illustrating the first-hand accounts by Kendall and Green of the Santa Fe and Mier expeditions, however, show some of the participants wearing felt hats with their uniforms rather than the prescribed military caps. Considering the weather extremes of Texas this is not surprising. What is surprising is that wide brimmed felt hats had no place officially in the uniforms of Texas.

Uniforms are portrayed in an oil painting owned by Pierce Butler of Nashville, Tennessee depicting what Time-Life Books, in *The Texans*, calls a conference between Texians and Cherokees in 1843.[152] In the lower right corner stands a military officer wearing a dark garrison cap, dark blue frock coat, light blue trousers with a yellow stripe and a sword. Squatting next to him is another White man wearing what appears to be a military cap and coat but with brown pants. He may or may not be military. In the center is an authoritative figure in what appears to be a military uniform, his coat open to reveal a red vest. To the right of center are five seated white men, three civilians, two in dark forage caps and outfitted in gray who may be military. Ordinarily paintings, even by contemporary artists, are highly suspect as far as authenticity is concerned. But this one was made by John Mix Stanley who actually attended the conference, according to Pierce Butler. Stanley gave the painting to a Pierce M. Butler who was an American official at the conference (the great-grandfather of the present owner of the painting). So its accuracy level may be considered higher than the usual historical art.

Are these Texas uniforms that should be included in this book? The blue ones have the identical cut of the American army uniforms of the day; but then so did some of the Texas uniforms. And there was often deviation from regulation whatever the case.

And what would American soldiers be doing in Texas in 1843 anyway – especially after the ruffled feathers from the Snively incident? Careful research uncovers the surprise that the blue uniforms are *American*, not Texian. What were they doing there? The answer, the contemporary Pierce Butler says, can be found in *The Texas Indian Papers*.

Pierce M. Butler, the Indian agent stationed in what is now Oklahoma, was sent to this annual powwow by the United States government with an escort of an officer and fifteen American soldiers, according to his namesake. The conference was held May 28 at Tehuacana Creek near Waco. The American in the red vest Butler believes to be his ancestor.[153] The Texas commissioners were General G.W. Terrell, Captain John S. Black and Thomas I. Smith. They may be the three White men in civilian clothing to the right of Butler. Indian tribes represented were Delaware, Shawnee, Caddo, Ioni, Aradarko, Tawakoni, Wichita and Keechi. *There were no Cherokees.*

Terrell, head of the Texas delegation, introduced Butler as one of the "big captains" of the "Great Father" of the United States, come to "witness the council and help make peace between Texas and his red brothers." Butler, a former governor of South Carolina, mentioned the American soldiers in his address. "My friend Captain Blake of the United States Army came along with me from the United States as an escort with fifteen men as authority from the government of the United States."[154] The officer in the lower right corner then is undoubtedly Captain Blake of the United States Army, not a Texian.

But as the council was held on Texas soil, Texas soldiers would have unquestionably been present also. The two White men in the military caps and gray tailcoats and trousers, it can be assumed with reasonable confidence, are Texas soldiers wearing the infantry fatigue uniform shown in Plate 9.

During 1843 the still smartly uniformed Texas Navy fought its two greatest battles: one against English-officered Mexican warships, the other against its most implacable enemy, Sam Houston. Houston's idea of economy, according to the *Handbook of Texas*, was to withhold all naval appropriations made by Congress.[155]

Now he went one step further. Abetted by a subservient Congress as short-sighted as himself as to the need for naval defense, he instigated an extraordinary secret session which voted to disband the navy and sell off its ships. This was particularly tragic, as Moore had only recently managed to renew the agreement with Yucatan, which had lapsed some time before.

Houston sent three commissioners to New Orleans, where Moore and his fleet were preparing for a return to the coast of Yucatan, to inform the commodore that he no longer had a navy to command. But Moore convinced the commissioners of the folly of the order. He sailed off with their blessing and with the head of the commission, Colonel James Morgan, aboard – ready to share the responsibility of defying Houston. When he heard this, Houston flew into a fury and issued a proclamation declaring Moore and his crews to be pirates, and to be dealt with as such by all nations. The order violated Texas law, which required a court martial before an officer could be dismissed.

In the meantime, before learning of the pirate proclamation, Moore and the Texas Navy won their greatest victory at sea in a running engagement with a Mexican flotilla that included two new steamers, *Guadalupe* and *Montezuma*, one an iron-clad. It was the first test between sail and steam. The Texians won, losing only five seamen while the Mexicans lost, by their own account, seventy-eight killed and even more wounded.

When he learned of Houston's vindictive order, Moore turned his ships back to Texas. They anchored at Galveston July 14. It was to be the last voyage of the Texas Navy, but the damage to the Mexican navy ships in that last battle successfully prevented any further invasion by sea from Mexico.

Moore had been able to supply his mariners with uniforms and other supplies by means of his personal credit in New Orleans when necessary, whereas Houston's personal credit was dishonored. Not the least of Houston's bitter hate for the Commodore might be attributed to that.

Moore proceeded to the capital at Austin for a showdown with the president. Houston handed him a dishonorable discharge. Moore countered by demanding a court martial, which was his right under Texas law.

In the meantime, Yucatan made peace with Mexico and the rental income from the navy was lost. Arrangements were made to auction the ships. But the auction was prevented by "forceful means" taken by the citizens of Galveston. It was an illusionary reprieve, however, for Houston and his ally Dr. Anson Jones, the next president, saw to it that the navy never left port again.[156]

1844

In August of 1844 Commodore Moore finally got his court martial. Houston stacked the deck by appointing as the judges three of his close personal friends. Even so, Moore was vindicated, even praised by the court. Next, a shame-faced Congress added to Houston's humiliation by passing resolutions praising Moore. In a last unbecoming act of petulance, Houston vetoed them. Congress passed them again over his veto.

In September Dr. Anson Jones was elected president. He took office December 9. He first appointed G.W. Hill as Secretary of War and Marine; then quickly replaced him with William G. Cooke. Cooke, who had arrived in Texas as an officer of the first uniformed unit to come to Texas at the beginning of the revolution in 1835, the New Orleans Grays, was, appropriately, to be the last to be in charge of the military forces of the republic.

1845

The uniform of the militia of New Braunfels reflected the Teutonic heritage of its settlers. This militia was organized from Germans settled at New Braunfels in 1845 by an Austrian-born nobleman, Prince Carl of Solms-Braunfels. The town was named after the Prince's estate in Europe. Prince Carl was Commissioner General for a colonization society, the Adelsverein, which brought many Germans to Texas. At New Braunfels the immigrants first built a stockade, the Zinkenburg, for protection from the Indians. Later, they constructed a blockhouse, the Sophienburg, named after the Prince's fiancee, Princess Sophia Salm-Salm. No trace of either remains, but the Sophienburg Museum, which perpetuates the memory of those early days, was built on the site of the blockhouse.[157]

★ ★ ★

Plate 26 (page 58)

PRINCE CARL OF SOLM-BRAUNFELS
New Braunfels Militia

Prince Carl's militia was described by four contemporaries: his successor to leadership of the Adelsverein, Baron John Meusebach, Herman Selle, M. Maris, a Frenchman who wrote *Souvenier d' Amerique*, and Fritz Goldbeck, one of the early settlers.

The Baron said they wore "hats bedecked with a cockade of rooster feathers, gauntlet gloves, and long clanking sabres."[158]

Selle remembered their uniforms as "long riding boots, gray woolen blouses, black velvet collars decorated with brass buttons, broad-brimmed hats trimmed with long black feathers, tilted back on their heads, swords buckled on, and armed with rifles."[159]

A translation of Maris' description in *Souvenier d' Amerique* by Yamile Dewailly, teacher of French at Austin Community College, says "the uniform consisted of high boots, gray pantaloons, gray blouse, and large white, broad-brimmed hat mounted by wild turkey feathers. Their arms consisted of a large cavalry sabre, a brace of horse pistols and a carbine 'Auswander in Texas'."[160]

Supposedly there is a jäger rifle in the Witte Museum in San Antonio inscribed on its barrel "for the defense of the German settlers in Texas against the Indians." However, when a special trip was made to inspect it there, the management of the museum could not locate it.

Goldbeck, a fourteen-year-old boy with the original settlers, remembered "dark gray uniforms, swords swinging at their sides."[161]

There are a few contradictions in the descriptions, such as what kind of feathers decorated their hats, and one said that they had *rifles* while another said *carbines*. Nevertheless, the descriptions are more than adequate for reconstructing the appearance of the New Braunfels militia.

Since Prince Solms had been an officer in the Austrian army the blouse would have probably been patterned after an Austrian military coat, according to Sam Nesmith, former Curator of the Alamo and researcher for the Institute of Texan Cultures, or possibly a vertically-pleated hunting frock like that shown in the only painting of the prince in which he is not wearing a metal breastplate. In that painting Solm's hunting frock is green or a grayish-brown. But his costume in that painting is *not* the militia uniform, though in some ways similar. Examine the background and you will see on a towering peak behind him a huge European castle-like structure of many rooms and stories. Undoubtedly the painting was made in Europe before the Prince came to Texas. It may well be, though, that the gray uniform of the New Braunfels militia was of quite similar cut, except for the collar and other minor details.[162]

★ ★ ★

The on-again, off-again annexation romance between Texas and the United States was now about to become a marriage. In a last gasp attempt to head it off, the Mexican government in late April grudgingly signed a preliminary agreement for a treaty of peace which would recognize the independence of Texas and settle other differences. But the sentiment for annexation was too strong to be stopped. In June the Texas senate approved plans for a convention of representatives of the people to vote on the annexation proposal offered by the United States. The convention met July 4 and approved the American offer overwhelmingly.[163]

Though the annexation would not be confirmed until the United States Congress approved a Texas state constitution, the U.S. Army now began to move into Texas. The Army of Occupation, as it was called, made its main camp on the north bank of the Nueces River, about where Corpus Christi is now. Under a strict agreement with President Jones, it was not to occupy the territory between the Nueces and the Rio Grande, which was disputed between Texas and Mexico. From this camp United States dragoons fanned out thinly across Texas taking over many of the defense functions of the Texas Army and rangers.

Except for Cooke, still sitting in his office in Austin, the Texas Army seems to have mysteriously evaporated during those last few months of 1845.

Writing to the president on December 12, Cooke advised that he was disbanding the Milam and Robertson County ranging companies when their enlistments expired in mid-month. Newly arrived American dragoons would be their replacements. The renewal of the three-months enlistments of other ranging companies would be subject to the approval of the U.S. Army authorities.[164] A great part of the euphoria over annexation was the naive expectation that the American takeover would relieve the Texians of any further need for self-defense, even rangers.

On December 29, the Congress of the United States approved the state constitution offered by Texas. All was formalized now except for the final ceremony transferring sovereignty.

1846

On February 19, 1846, the final act in the dismantlement of the Republic of Texas would be performed at Austin, the capital, a hamlet of only 250 permanent residents. But to the Texians it was the city of progress envisioned by Lamar, who chose the site. Coming to town after a long absence, veteran ranger Ben Gooch was impressed. "I had a splendid horse, he could keep his feet anywhere ... I didn't hesitate to run him over the rockiest of country. He was the surest-footed animal I ever saw. Well Sir, I rode that horse down to Austin, and what do you think? They had built so many houses on the avenue and hung out so many signs that that horse was kept so busy looking at them, blame my skin if he didn't fall down three times on the level avenue."[165]

Quite a throng was now converging on Austin to witness the final drama. At high noon officials of the Republic of Texas and the State of Texas took their seats on the long gallery of the east side of the capitol. They were escorted by uniformed officers of the U.S. Army already stationed at Austin.

As a member of the outgoing president's cabinet, William Gordon Cooke was entitled to sit on the porch with the assembled dignitaries. He may have been the only person present wearing the uniform of the Republic of Texas. If so, considering the solemnity of the occasion it would have been the full-dress one, without doubt. His colonel's uniform would have been like the captain shown in Plate 7, except the collar embroidery would have been a silver live oak wreath on each side. Cooke had answered the Texian appeal for help in 1835 as a twenty-seven-year-old captain in the New Orleans Grays, the first uniformed unit of the revolution. In the Texas Army he had risen to the rank of colonel. It was Cooke who had laid out the military road to connect the various military posts; who had accompanied the Santa Fe Expedition and been captured and imprisoned in Mexico with its survivors; and served as Quartermaster General; was also an officer in the Texas Marines; and now, in his final role, was Secretary of War and Marine. During the last several months he had watched as his army faded away as the United States Army moved in and replaced it. He was now literally the last man to wear a uniform of the Republic of Texas.

If uniformed soldiers or sailors of the republic were also included in the honor guard on this occasion, it was not mentioned by observers. But many were there, besides Cooke, who had worn proudly the uniforms of the Republic of Texas and the Texas Revolution. Most were young and still in the prime of life like Cooke; after all, only ten years had elapsed from the beginning of the revolution to the end of the republic. It had been ten years of constant warfare and adventures, both glorious and disastrous. Present were veterans of Bexar, San Jacinto, the Council House Fight and other Indian battles, Santa Fe and Mier, as well as engagements at sea. It was not a light thing for them to see the lone star banner they had served come down. "There was a smothered sensation which all felt, yet few desired to display in public," a newspaperman present noticed. "Broad chests heaved – strong hands were clinched, and tears flowed down cheeks where they had been strangers for long years."[166]

Anson Jones, last president of the Republic of Texas, now gave a short valediction. Described as "appropriate, eloquent, and at times pathetic,"[167] by Frank Brown, then a thirteen-year-old boy among the spectators, it ended with the words: "The final act in the great drama is ended. The Republic of Texas is no more."[168]

The lone star flag was lowered. The stars and stripes went up. A cannon salute by United States artillerymen then sounded the death knell of the Republic of Texas, and with it the faded dream of the visionary Mirabeau Buonaparte Lamar and others that it would some day become an empire to rival the United States.

NOTES

1. *Texas Citizen Soldiers*, bulletin by Star of the Republic Museum (n.d.); Nance, *After San Jacinto*, 33.
2. Forbes purchase in Army Papers, Texas State Archives, box 1218-28. Toby shipments in Jenkins, ed., *Papers*, VII, 467 (item 3697); 407 (item 3721); 507 (item 3752); VIII, 207 (item 3962), 216-217 (item 3973). Forbes order VI, 288 (item 3059).
3. Ibid.
4. Ibid., VIV, 343-344 (item 4365); Gammel, *Laws of Texas 1822-1897*, I, 93 (item 997).
5. Koury, *Arms for Texas*, 14-17.
6. Gilbert, *Concise History of Texas*, 58-59.
7. Shelton, *Uniform Buttons*, 58-59.
8. Wortham, *History of Texas*, III, 365.
9. Gray, *Diary*, 217.
10. Nance, *After San Jacinto*, 31.
11. Bate, *General Sidney Sherman*, 159.
12. Ibid., 156.
13. Nance, *After San Jacinto*, 17.
14. Prints and Photographs Collection. Center for American History, UT-Austin.
15. Herman Warner Williams, Director, Corcoran Museum, to J. Hefter, Instituto Internacional de Historia Militar, August 10, 1970; Hefter to Williams, July 30, 1970; Marisa Kelly, Acting Archivist, Corcoran Museum, to author, July 11, 1944. Ed Milligan to author, June 17, 1994; all in author's possession.
16. Ibid.
17. Gilbert, *Concise History of Texas*, 60-61.
18. Nance, *After San Jacinto*, 28.
19. Army Papers, Texas State Archives, box 1228, folder 27.
20. Ibid.
21. Ibid., 1228-28.
22. Ibid.
23. Ibid., 1228-29.
24. Smithwick, *Evolution of a State*, 158-159.
25. "Buttons of the Republic of Texas," *National Button Bulletin*, National Button Society, Akron: No. 2, May 1986.
26. Army Papers, Texas State Archives, 1837, 1838, 1839, 1840, folder 4.
27. Ibid., box 1228, folder 35.
28. Ibid., 1217-5.
29. Wright, Thomas Jefferson. Oil portrait of Juan N. Seguín, Legislative Reference Library, Texas State Capitol, Austin.
30. General Order Number 5, Army Papers, Texas State Archives, 6-10; Hefter, Army, Plate IV.
31. Stiff, *Texas Emigrant*, 87.
32. Sheridan, *Galveston Island*, 112.
33. General Order Number 5, Army Papers, Texas State Archives, 4-6; Hefter, *Army*, Plates III and VI.
34. Stiff, *Texas Emigrant*, 71-72; Hogan, *Texas Republic*, 49.
35. Siegel, *Political History*, 103.
36. General Order Number 5, Army Papers, Texas State Archives, preamble. All other histories show Johnston's first name as Sidney. However, on this order his signature is spelled Sydney.
37. General Order No. 5, Army Papers, Texas State Archives, 4-5; Hefter, *Army*, Plate III.
38. Ibid., 5-6; Hefter, *Army*, Plate IV.
39. Army Papers, Texas State Archives, 401-1267/7, 401-1228/16, Cressman contract.
40. Koury, *Arms for Texas*, 14.
41. General Order No. 5, Army Papers, Texas State Archives, 5-6; Hefter, *Army*, Plate IV.
42. Ibid.
43. Uniform display, San Jacinto Monument, La Porte, Texas.
44. Ibid.
45. Brian Butcher, Director of Research, San Jacinto-Museum Association, to author, May 25, 1995. Letter in author's possession.
46. Army Papers, Texas State Archives, box 1224-2; General Order No. 5, 6-7.
47. General Order No. 5, Army Paper, Texas State Archives, 6-10; Hefter. *Army*, Plate IV.
48. Koury, *Arms for Texas*, 62; Nance, *After San Jacinto*, 90-91.
49. Hefter, *Army*, Plate IV; Jenkins, ed., *Papers*, VIII, 148-149 (item 3947).
50. General Order No. 5, Texas State Archives, 13; Hefter, *Army*, Plate IV.
51. Ibid.,8-10; Hefter, *Army*, Plate IV; Steffen, *The Horse Soldier*, Uniform Color Plate I.
52. Koury, *Arms for Texas*, 65.
53. Pierce, *Texas Under Arms*, 130-131.
54. General Order No. 5, Army Papers, Texas State Archives, 8; Hefter, *Army*, Plate II.
55. Ibid.
56. Ibid.
57. Ibid., 13.
58. Regulations and Instructions for the Government of the Naval Service of Texas, December 15, 1836. Texas State Archives, 43, Article 10.
59. Naval General Order 13th March 1839, Texas State Archives, 1; Hefter, Navy, Plates 1, 2, 6.
60. Ibid.
61. Ibid.
62. Matthews, *Ten Battle Flags*, text and illustration 10 for Hawkins flag; Hefter, *Navy*, Plates 1 and 2 for Variant flag.
63. Naval General Order, March 13, 1839, Texas State Archives, 1; Hefter, *Navy*, Plates 1, 2, 3, 6.
64. Koury, *Army for Texas*, 36.
65. Ibid., 35-36.
66. Naval General Order, March 13, 1839, Texas State Archives, 1; Hefter, *Navy*, Plates 1, 3, 6.
67. Garland Roark to author, April 13, 1973, letter in author's possession; Ed Milligan, *Texas Republic Navy*, CMH Plate 395; numerous "hands of grass" entries in naval and ships inventories.
68. Sheridan, *Galveston Island*, 77-78.
69. Naval General Order, March 13, 1839, 2, and Regulations (naval), December 15, 1836, 43, Article 10, Texas State Archives; Hefter, *Navy*, Plate 1.
70. Jenkins, "the Texas Navy," *the Republic of Texas*, 35.
71. Regulations (naval), December 15, 1836, Texas State Archives, 43, Article 10; Hefter, *Navy*, Plates 1 and 3.
72. *Texas Navy in Yucatan, 1837*, Rosenberg Library Bulletin N.S. Vol. 6, No. 1, March, 1976, 4, 7.
73. Ibid., (n. 4).
74. Naval General Order, March 13, 1839, Texas State Archives, 2.
75. Johns (Edward) Papers 1841-1845, Ships Journals, Texas Navy Papers, Center for American History, UT-Austin.
76. Navy Papers, Texas State Archives, Return of Slops and Small Stores of the Austin, year ending December 13, 1840.
77. Jenkins, "the Texas Navy," *the Republic of Texas*, 35; Gray, "Campeche was Texas Navy's Last Engagement," Port of Galveston Bicentennial Appointment Calendar 1976, original sketch, 54; photo of 1841 watercolor *Texas Sloop of War Austin*, in author's possession.
78. J. Hefter to author, December 13, 1972, note in author's possession.
79. Hefter, *Navy*, Plate 4.
80. Ibid.
81. Koury, *Arms for Texas*, 44.
82. Jenkins, "The Texas Navy," *The Republic of Texas*, 35.
83. Hill, *The Texas Navy*, 20.
84. Wells, *Commodore Moore and the Texas Navy*, 17, 20, 53, 62, 102, 116.
85. Gray, Alfred G. "Campeche was Texas Navy's Last Engagement," *Port of Galveston Bicentennial Appointment Calendar and Compendium, 1976*, 55.
86. Army Papers, Texas State Archives, box 401-1267, folder 6. John R. Burke contract, January 23, 1840.
87. Ibid., 401-1267/7; 401-1228/16. William Cressman contract, November 1, 1839.
88. Ibid., Tryon contract; Koury, *Arms for Texas*, 14-17, 85.
89. Ibid., 401-1267/6. John Malseed contract, November 1, 1839.
90. Ibid., Gilbert Cleland contract, November 20, 1839.
91. Ibid., 401-1228/39. Magee, Taber (or Faben) & Company contract, November 16, 1839.
92. Ibid., 1228-16 and 1228-39. Lutz (or Lutts) contract, November 19, 1839.

93. Ibid., box 401-1217/folders 6, 7, 8, 9, 10, 15; and 1228/18, 1228/39. Various certificates of inspection: November 20, 1839, inspector's signature semi-legible (appears to be P. Crelando), 840 forage caps of blue cloth (undoubtedly the Gilbert Cleland contract); December 20, 1839, same inspector, 280 black leather cavalry belts and brass plates for the Texas Army at the store of (semi-legible, appears to be M.C. Hurstoram), 8 Maiden Lane (New York, New York). These are not the Cressman contract, which was for *infantry* belts with *white metal* plates; December 31, 1839, M. Grier, U.S. Inspector, Philadelphia, 840 wooden canteens (the Henry Lutz contract); January 20, 1840, H.H. Jones, New York, for 280 American drill jackets, 560 American drill pants, 280 uniform coats, 280 uniform pants, 560 infantry drill jackets, 1,120 infantry drill pants, 560 uniform infantry coats, 560 uniform infantry pants, 840 haversacks, 2,240 cotton shirts, 30 American gray jackets, 125 American gray pants, 40 infantry gray jackets, 200 infantry gray pants (probably the bulk of the Burke contract); March 2, 1840, Jonas Gurnee, New York, "clothing made by John R. Burke for government of Texas", 908 gray infantry pantaloons, 438 gray infantry jackets, 404 gray cavalry pantaloons, 236 gray cavalry jackets.

94. Ibid., 401-1267/6. Abstract of Costs incurred by Daingerfield (n.d.), including the saddle for Colonel Wells.

95. Nance, *After San Jacinto*, 214.

96. Army Papers, Texas State Archives, 401-1224/19, inventory. Also 401-1267/7, Burke order shipped on *Metron*.

97. Ibid., *Harriet Porter* cargo list.

98. Ibid., 401-1217/16.

99. Ibid., Treasury Department, Auditors Office records, Settlement of Accounts by Charles Mason, First Auditor, with W.G. Cooke, Quartermaster General, by quarter ending 31, March 1840.

100. Maverick, *Memoirs*, 31-44.

101. Ibid.; Brown, *History of Texas*, II, 175-177; Yoakum, *History of Texas*, II, 298-299; Johnson, *History of Texas and Texans*, I, 463-464 (casualty figures taken from official report of Hugh McLeod, the Adjutant and Inspector General, in Johnson).

102. Koury, *Arms for Texas*, 40.

103. Weems, *Dreams of Empire*, 175.

104. Stiff, *Texas Emigrant*, 69.

105. Army Papers, Texas State Archives, Cooke to Neighbors, June 6, 1840, re: Border Guards, "Furnish them with quartermaster stores and ordnance." The Guards were a volunteer company from Galveston and the stores referred to were there.

106. Treasury Department, Office of the First Auditor records, July 26, 1840, Audit of Accounts of William G. Cooke by Charles Mason, First Auditor, Texas State Archives.

107. Major General Thomas S. Bishop talk before Sons of Confederate Veterans, Austin, Texas, May, 1986.

108. Jenkins, *Recollections of Early Texas*, 64-65.

109. Nance, *After San Jacinto*, 343, 356-357.

110. Ibid., 358.

111. Army Papers, Texas State Archives, 401-1241/1 (oversized documents). Exhibit of Monies Received and Disbursed by William G. Cooke, Quartermaster General, from 1 October 1839 to 30 September 1840.

112. Ibid., 1217-25.

113. Ibid., 401-1267/7. List of army suppliers and one inspector (Grier).

114. Navy Papers, Texas State Archives, Naval Storekeeper Public Stores and Navy Yard inventory list, January 1840.

115. Ibid., 401-1240/8. July 1, 1840, quarterly return, Public Stores at the Texas Stores and Navy Yard, Galveston.

116. Ibid., Slops and Small Stores for quarter June 12, 1840 to September 30, 1840.

117. Ibid., 401-1240/3 (oversized documents). Return of Slops and Small Stores, the *San Antonio*, Alexander Moore, lieutenant commanding, for quarter ending 30 September, 1840.

118. Ibid., 401-1240/1 (oversized documents). Return of Slops, etc., for the *Austin*, year ending December 13, 1840.

119. Army correspondence, Texas State Archives, 401-1307.

120. Santa Fe Expedition papers, Texas State Archives, box 25, folder 14 (oversized documents).

121. Kendall, *Narrative of an Expedition*, I, 315.

122. Ibid., I, 406-417.

123. Ibid., I, 423; II, 56.

124. Koury, *Arms for Texas*, 14.

125. Ibid., 53.

126. Ibid., 35.

127. Kendall, *Narrative of an Expedition*, I, 304.

128. Navy Papers, Texas State Archives, January 1, 1841 inventory list; plus Statement of Receipts and Issues of John G. Tod, late Naval Storekeeper (n.d.).

129. Ibid., folder 14, Other Financial Records - Navy. Abstract of Purchases made by Commodore E.W. Moore, Navy, for quarter ending June 30, 1841.

130. Army Papers, Texas State Archives, 401-1224/1.

131. Bollaert, *William Bollaert's Texas*, 45-51.

132. Ibid.

133. Ibid.

134. Ibid.

135. Ibid.

136. Ibid.

137. Ibid., 39; Koury, *Arms for Texas*, 53-57.

138. Koury, *Arms for Texas*, 56.

139. Army Papers, Texas State Archives, 401-1224/17.

140. Ibid., 401-1226/10.

141. Ibid.

142. Ibid., 401-1224/3.

143. Ibid., April 27, 1842 report by W.J. Mills, Inspecting and Mustering Officer, of supplies in possession of volunteers mustered into service of Republic of Texas.

144. Ibid., Quartermaster correspondence, November 8, 1842, J. Snively to Colonel W.C. Washington, saying the volunteer army under General Davis was supplied out of the stores on hand at Galveston; Bollaert, *William Bollaert's Texas*, 119; Yoakum, *History of Texas*, II, 362.

145. Sanchez Lamego, *Siege and Taking of the Alamo*, 46.

146. Yoakum, *History of Texas*, II, 365.

147. Green, *Journal of the Texian Expedition against Mier*, 45-47.

148. Ibid.

149. Ibid., 63, 71; Scott, *History of Coryell County*, 59.

150. Army Papers, Texas State Archives, 401-1308/11. M.C. Hamilton, Acting Secretary of War and Marine, to Colonel Jacob Snively, January 28, 1843.

151. Ibid., 401-1224/7. Articles received by Major J.W. Tinsley from M.J. Warfield in third quarter of 1843. Nevin, *The Texans*, 169.

152. Nevin, *The Texans*, 169.

153. Butler, Pierce, two telephone interviews by author, January, 1995; letter of same month in author's possession.

154. Winfrey and Day, *Texas Indian Papers*, I, 156-157.

155. Daniel, James M. "Texas Navy," *Handbook of Texas*, II, 749-751.

156. Oates, ed., *Republic of Texas*, 40-41; *Handbook of Texas*, Webb, ed., II, 749-751.

157. Webb, ed., *Handbook of Texas*, II, 635, 638.

158. Sam Nesmith, former curator of the Alamo and former researcher for the University of Texas Institute of Texan Cultures at San Antonio (Muesbach quote).

159. Ibid., (Selle quote).

160. Ibid., (Maris quote).

161. Pierce, *Texas Under Arms*, 114.

162. Oil painting of Prince Solm in Nevin, *The Texans*, 161.

163. Wortham, *History of Texas*, IIII, 206.

164. Army Papers, Texas State Archives, 401-1215/42/43; Army Correspondence, William G. Cooke, Secretary of War and Marine, to President Anson Jones, January 10, 1846.

165. Brown, *Annals of Travis County*, V and VI, 13.

166. De Shields, *They Sat in High Places*, 170.

167. Brown, *Annals of Travis County*, V and VI, 19.

168. Ibid.; Nevin, *The Texans*, 219.

Republic of Texas Army
infantry - 1836

T E X
©Bruce Marshall KEWE
Republic of Texas
one of several early variations
Army 1836

Bruce Marshall
©
KCWE
Republic of Texas
Felix Huston (fatigue dress)
brigadier general/army

© Bruce Marshall KCWE
Republic of Texas
Dragoons-1836

Republic of Texas
Texas Rangers
Coleman's Fort-1837

Republic of Texas
Lt. Col. Juan Seguin
Cavalry 1838

Republic of Texas
Army captain
full dress uniform
Milam Guards

©
Bruce Marshall
KCWC
Republic of Texas
Army
the Milam Guards full dress uniform
sergeant

© Bruce Marshall, KCWE
Republic of Texas
Army
sergeant fatigue uniform
infantry

Bruce Marshall
©
KCWR
Republic of Texas
Thomas Jefferson Chambers
major general

1st CAVALRY
TA
© Bruce Marshall, KCWE
Republic of Texas
1842 / Sergeant / full dress
Cavalry
the Travis Guards

Republic of Texas
trumpeter
Cavalry

Republic of Texas
cavalry
officer/summer undress uniform

1st CAVALRY
Bruce Marshall ©
Republic of Texas
fatigue dress
Cavalry
corporal

Republic of Texas
summer fatigue uniform
Cavalry
(cavalry flag/cavalry guidon)

© Bruce Marshall KCWE
Republic of Texas
Captain/Navy
in full dress

© Bruce Marshall KCWE
Republic of Texas
Navy/captain
summer undress uniform

© Bruce Marshall
Republic of Texas
summer uniform
Navy / midshipman
Hawkins naval flag / Wharton ship's ensign

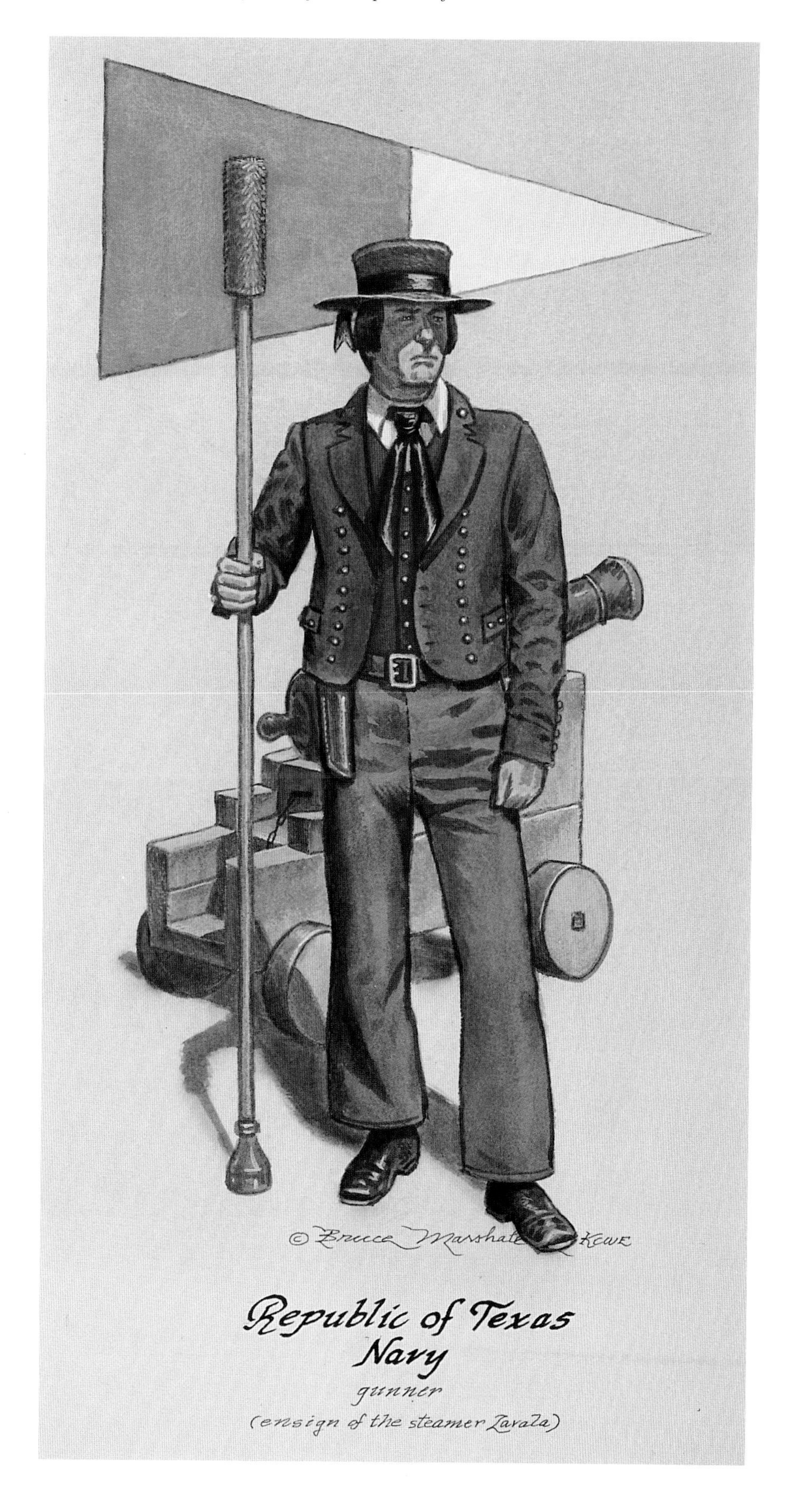

Republic of Texas
Navy
gunner
(ensign of the steamer Zavala)

© Bruce Marshall KCWE
Republic of Texas
Navy
summer whites

© Bruce Marshall
KCWE
Republic of Texas
foul weather gear
Navy
(ensign of flagship Austin)

©
Bruce Marshall
Republic of Texas
Marine Lieutenant

© Bruce Marshall, KEWF
Republic of Texas
Marine Corps
sergent/fatigue uniform

©Bruce Marshall, KCWE
Republic of Texas
(service dress~U.S. surplus 1834)
Marine Corps
(ensign of the ship San Antonio)

Republic of Texas
Galveston Coast Guards

Bruce Marshall
©
KCWE
Republic of Texas
Prince Carl of Solms-Braunfels
New Braunfels militia

APPENDICES

APPENDIX I
Army of the Republic of Texas uniform regulations
facsimile reproduction

UNIFORM OF THE ARMY

OF THE

REPUBLIC OF TEXAS,

PRESCRIBED AND PUBLISHED

BY ORDER OF THE PRESIDENT.

ADJUTANT GENERAL'S OFFICE,
Houston, May 23, 1839.

GENERAL ORDER,
No. 5

The President has been pleased to adopt the present uniform for the army of the Republic of Texas. It will be the duty of the officers and men to adhere rigidly to it.

By order of the Secretary of War

H. McLEOD,
Adjutant General.

Albert Sydney Johnston

UNIFORM, &c.

MAJOR GENERAL.—*Coat*—dark blue cloth, stand-up collar, not higher than the chin, with an embroidered leaf of live oak, and a star on each side; double breasted, one row of buttons on each side, ten in each row, in clusters of three, and one at the top, the rows to be five inches apart at top, and three at the bottom; slash sleeves, four small buttons; pointed cross flaps to the skirts, with four buttons, two in the centre, and one at each end of the flaps, the skirt to reach within three inches of the bend of the knee, with turn backs of cloth, the bottom of the skirt not less than four nor more than five inches broad, with a gold embroidered star on the tail of each skirt, two inches in diameter; two hip buttons to range with the lower buttons on the breast; lining with black silk; two buttons on the fold of each skirt, and one at the tail of each skirt.

Trousers—dark blue cloth, with one stripe of buff cloth, one inch wide, up each outward seam, welted at the edges; plain white linen for summer.

Vest—for winter, buff, double breasted, with small uniform buttons; white for summer.

Epaulettes—gold, with solid crescent; a star one inch and a half in diameter in the crescent; dead and bright gold bullion.

Buttons—according to pattern.

Hat—cocked, without binding; fan or back part not more than ten nor less than eight inches wide; the front, or cock, not more than eight, nor less than five inches; each corner five inches; black ribbons on the two front sides.

UNIFORM OF THE ARMY.

Uniform, or Dress of the Army.

Loop and cockade—cockade of black silk; loop gold six inches long.

Stock—black satin or patent leather.

Boots—leg.

Spurs—Gilt.

Sword and Scabbard—according to pattern.

Sword-knot—gold braid with bullion hangers.

Gloves—black for winter, white for summer.

Sword-belt—black patent leather; carriages of the same material, the *belt* to be worn over the coat.

Plate—gilt, having a star and a sprig of live oak on each side in gold.

Sash—yellow silk net, with fringe ends to go twice round the waist and tie on the right hip.

FOR BRIGADIER GENERAL.—*Coat*—same as for Major General, except the buttons to be ranged in clusters of two on the breast.

Trousers—same as for Major General.

Vest— do. do do.

Sash—red; all other equipments same as for Major General.

UNDRESS.

Coat—citizens', with rolling collar, two rows of buttons arranged as on the uniform coat, four on the flaps, one at each hip, and one on each skirt.

Frock Coat—dark blue cloth, double breasted, two rows of buttons placed according to rank, as on the dress coat; stand-up collar; lining black silk; pockets in the folds of the skirts, with one button at the hip, two in each fold, and one at the end of each pocket.

Trousers—same as full dress.

Cloak—blue cloth, lined with scarlet marino.

Uniform of Officers of the General Staff.

Coat—corresponding with that of general officers, without the star on the skirt, and with the buttons at equal distances, ten in each row.

Trousers—same as for general officer.

Vest— do do do

UNIFORM OF THE ARMY.

Uniform, or Dress of the Army.

Epaulettes—same as for general officer without the star.

Sash—yellow silk net, to go twice round the waist and tie on the right hip.

Hat Loop and Cockade *Sword and Scabbard* *Belt and Plate* *Sword-knot* *Stock* *Boots* *Gloves* *Cloak*	The same as for general officers.

Aguillette—twisted gold cord, with gilt engraved tags.

UNDRESS.

Coat—as described for general officers, except the buttons at equal distances.

Frock Coat—same as for general officers, except the buttons, which will be at equal distances.

Trousers—same as for general officers.

Pay Department.

Coat—dark blue cloth, double breasted; standing collar; two rows of buttons ten in each row at equal distances apart, to commence at the collar, and run in a right line to the bottom of the lappel: four inches apart at top, and two and a half at the bottom; skirts to be made after the style of the citizen's coat, with a button at each hip, and at the end of each fold, and one intermediate in each fold; buttons the same as for other officers of the general staff.

Trousers—same color as coat, but without ornament.

Hat—plain, of the same form and dimensions as described for general officers; black leather, black silk gilt loop, no tassel or other ornament except the cockade and star.

Sword—according to pattern.

Stock—black satin or patent leather.

Sword-knot *Belt and Plate* *Epaulettes* *Spurs* *Gloves*	Same as heretofore described.
Frock Coat *Cloak* *Vest*	Same as heretofore described, except the cloak to be lined with blue.

UNIFORM OF THE ARMY.

Uniform, or Dress of the Army.

Medical Department.

Coat—same as for pay department, except the collar, which will be of black velvet.

Trousers—same as described, with black stripe up each outward seam, one inch wide.

All other trimmings, equipments, &c., same as heretofore described except the sash.

Infantry Uniform.

FOR FIELD OFFICERS.—*Coat*—dark blue cloth, double breasted, two rows of buttons, ten in each row, to be placed at equal distances; the distance between the rows to be four inches at top, and two at the bottom; measuring from the centre or eye of the button, to curve outward from the centre of the breast; standing collar, to meet in front with hooks and eyes, to rise no higher than to permit the free turning of the chin, to be embroidered on each side with a silver live oak wreath, seven inches in length; plain round cuff, three inches deep, with four large buttons round the top, and to button with three small buttons at the under seam; four buttons at each flap; two on the hips, and one at the bottom of each skirt, the skirt to extend within three inches of the bend of the knee; turn backs to be of the same cloth as the coat, silver embroidered star of five points, at the bottom of each skirt, and lining to be of the same color as the coat.

FOR CAPTAINS—the same as field officers, with the exception of the collar, instead of embroidery, to be framed with half inch silver lace, two loops of quarter inch lace, four and a half inches long on each side of the collar, with one small uniform button at the end of each loop.

FOR LIEUTENANTS—the same as for Captains, with the exception of the collar, to have only one loop on each side.

Trousers for Field Officers and Commissioned Staff—same color as the coat, with two stripes of white cassimere three fourths of an inch wide, up each outward seam, leaving a light between, and welted at the edges.

FOR COMPANY OFFICERS—same as above, with but one

UNIFORM OF THE ARMY.

Uniform, or Dress of the Army.

stripe up each outward seam, one inch wide, the stripes to be in advance of the seam; plain white linen for summer.

Boots—Wellington.

Vest and Gloves—white.

Waist-belt and Stock—black patent leather.

Sword and Clasp—according to pattern.

Sword-knot—crimson and silver, with a bullion tassel.

Sash—crimson silk net, with silk bullion fringe ends, to go twice round the waist, and tie on the right hip, the pendant part to be uniformly one foot in length from the tie.

Cap—to be of beaver, bell crown, plated scales, silver star in front surrounded by rays, with number of regiment underneath, and the letter of the company within the star, the peak black patent leather, well depressed; socket for plume to be plated; tassels of silver cord and bullion.

Plume—to be of white cocks' feathers, falling from an upright stem.

Epaulettes—according to pattern. Captains to wear one on the right shoulder, and subalterns one on the left.

Aiguillettes—twisted silver cord, to be worn on the right shoulder under the epaulettes; worn only by the commissioned staff.

Buttons—according to pattern.

UNDRESS.

Frock Coat—dark blue cloth, single breasted, standing collar, to meet in front; one row of ten buttons at equal distances; plain sleeve with three small buttons, two large buttons on the hips, two at the folds of each pocket, and one at the bottom of each fold.

Trousers—dark blue cloth, with a white cassimere cord down the outer seam.

Forage Cap—blue cloth, according to pattern; cover of black patent oiled silk.

Cloak—dark blue cloth, lining blue serge.

Uniform for Non-commissioned Officers and Privates.

Coat—dark blue, single breasted, one row of ten buttons, standing collar, to meet in front; one loop of white worsted

UNIFORM OF THE ARMY.

Uniform, or Dress of the Army.

half inch lace four inches long on each side of the collar, with a button at the end of each; skirt of the coat to extend within three inches of the bend of the knee, a worsted star at the end of each skirt, two buttons on the hip, and one at the bottom of each skirt; shoulder straps of white worsted cord, to extend from the top of the shoulder to a button within one inch of the collar; four buttons round the cuff to open two inches at the under seam.

NON-COMMISSIONED STAFF — same as privates, with three chevrons on the right arm above the elbow, the points of the lower chevron joined by a concave curve.

FIRST-SERGEANTS—same as privates, with three chevrons on the right arm above the elbow, the points of the lower chevron joined by a straight line.
All other sergeants, same as above, omitting the straight line of chevrons.

CORPORALS—same as privates, with two chevrons below the right elbow.

Trousers—dark blue cloth, with a worsted stripe one inch wide, in front of the outward seam; white cotton drilling for summer.

Cap
Bootees } According to pattern.
Belt

Fatigue Dress.

Jacket—grey cloth, single breasted, one row of nine buttons in front, with the same distinction for non-commissioned officers as in full dress; the worsted lace to be black.

Trousers—grey cloth without ornament.

Great Coat—grey cloth, with cape and stand-up collar.

Forage Cap—according to pattern.

Knapsack— do. do.

Non-commissioned Staff and First Sergeants to wear red worsted sashes.

Uniform of Cavalry.

Coat—dark blue cloth, single breasted, three rows of buttons,

UNIFORM OF THE ARMY.

Uniform, or Dress of the Army.

nine in the front, and eleven in each of the outside rows, the outside rows to be five inches apart at top; and three inches at the bottom, to swell outwards with the breast towards the top, the buttons gilt, lace gold, the skirt not over four inches in length, two buttons at the bottom of the skirt, and two on the hips; the cuffs three inches deep, four buttons round the top, and to button with three small buttons at the under seam; the trimmings of the collar to designate rank, same as in infantry, except that the lace and embroidery will be gold instead of silver.

Trousers—Field-officers and commissioned staff—dark blue cloth, to button up in front, the buttons covered with a flap, one stripe of buff cloth up each outward seam, one and a half inches wide, and welted at the edges.

FOR COMPANY OFFICERS—same as above, except the stripe which will be but one inch wide, stripe to be in advance of the seam; for summer, plain white *trousers* cut after the same fashion as above.

Scales—all officers to wear scales on each shoulder, made according to pattern; the scales of field officers trimmed with gold bullion; captains to wear bullion on the scale of the right shoulder; lieutenants on the left.

Aiguillettes—twisted gold cord, with gilt tags worn on the right shoulder, under the scale, to be worn by field officers and commissioned staff.

Buttons—according to pattern.

Helmet—according to pattern, with white horse hair plume, the Colonel to be designed by a tuft of red horse hair in the front of his plume.

Boots—leg.

Spurs—yellow metal.

Sabre—according to pattern.

Knot—buff leather with tassels at the end.

Sash—crimson silk net, with a tassel at each end, to go twice round the waist and tie on the right hip, the pendant part to be eighteen inches in length from the tie. The sash may be worn with the uniform or undress coat, and on all occasions when the officer is mounted.

Sword-belt—black patent leather, two inches wide and made according to pattern.

Stock—black.

Gloves—bull leather.

UNIFORM OF THE ARMY.

Uniform, or Dress of the Army.

UNDRESS.

Coat—dark blue cloth, cut after the fashion of citizen's coat, with falling collar, two rows of buttons in front, eleven in each row, the rows to commence at the outer edge of the collar, and to be two inches apart at the bottom to swell outwards towards the top; slash sleeves, the slash to be six inches in length, and four buttons to each, three buttons along each flap, two on the hips, one at the bottom of each skirt, and one about midway of each skirt; shoulder strap on the shoulder.

This coat may be worn (with or without scales according to orders) upon all duty done by detail, when the officer is not required to be in full uniform, and upon all drills where the troops are not in full dress.

Officers upon stable duty, or active service, may be permitted to wear shell jackets, corresponding in cut and color with that of the men, and trimmed with black braid, according to pattern; straps on the shoulder.

Trousers—same as in full dress; plain white for summer.

Officers on stable duty, or active service, will be permitted to wear overalls, corresponding in color with that of the coat, or jacket, to be made full, and to button up to the point of the hip, on each side, the buttons one and a half inches apart, and covered with a flap. All trousers and overalls (excepting those worn with the undress coat) to be foxed with cloth of the same material according to pattern.

Vest—buff, double breasted, with small uniform buttons.

Cloak—dark blue cloth, lined with scarlet, stand-up collar, cape to reach to the bottom of the coat sleeve, arm holes under the cape; when on active service a belt may be worn of the same material as the cloak, two and a half inches wide, and to button with three buttons in front.

Forage Cap—according to pattern.

Uniform of Non-commissioned Officers, Buglers, and Privates of Cavalry.

Coat—dark blue cloth, cut after the same fashion as the commissioned officers, buttons same as to number and position as the officers' coats; brass scales on the shoulder. The collar of

UNIFORM OF THE ARMY.

Uniform, or Dress of the Army.

Sergeants and Chief Musicians' coats, to be trimmed with yellow worsted binding, after the style of the Lieutenants. Those of privates to be framed with yellow worsted lace. Musician's coat to be of scarlet cloth. Sergeants to wear chevrons of three bars on each sleeve above the elbow, points towards the cuffs. Corporals, two bars below the elbow. Lance-Corporals one bar. Non-commissioned staff to wear yellow worsted bullion on the scale of the left shoulder.

Trousers—same color and cut, and made after the same fashion as the officers. Sergeants to have one yellow stripe three-fourths of an inch wide, up each outward seam. Corporals and Privates a stripe half an inch wide, the stripe to be in advance of the seam. White cotton drilling for summer.

All cavalry trousers to be foxed with cloth of same material.

Helmet—according to pattern.

Gloves—buff leather.

Boots—ankle.

Non-commissioned Staff, First Sergeants of Troops, and Musicians, to wear red worsted sashes.

Fatigue Dress.

Jacket—cadet grey cloth, stand-up collar, single breasted, one row of ten buttons in front, and trimmed according to pattern. Non-commissioned officers, designated the same as in full dress; chevrons to be of black braid; white cotton drilling for summer, made after the same fashion as above.

Jackets to be worn, with or without scales, according to order.

Trousers—same color as the jacket, and the same fashion as the full dress, without the stripes; white cotton drilling for summer.

Forage Cap—according to pattern.

Great Coat—grey cloth, stand-up collar, single breasted, cape to reach down to the cuffs, and to button all the way up.

Arms and Equipments for Cavalry.

Sabre, *Carbine*, *Pistol* } According to pattern in Ordnance Department.

Sword belt, *Cartridge box*, *Pouch, &c.* } Black leather, and made according to pattern.

UNIFORM OF THE ARMY.

Uniform, or Dress of the Army.

Saddles—according to pattern.

Bridle
Collar
Halter
Saddle bags } Black leather, and made according to pattern.

Holsters—with black leather covers.

Haversack
Forage bag } According to pattern.

Spurs—yellow metal.

Curry Comb
Brush } According to pattern.

Ordnance Uniform.

Coat—dark blue cloth, double breasted, two rows of buttons ten in each row, at equal distances, the rows to be four inches apart at top, and two at the bottom, measuring from the centre or eye of the button; stand-up collar, to meet in front with hooks and eyes, and to rise no higher than to permit the free turning of the chin; collar edged with buff cloth, and braided in gold with oak leaves, and acorns and a grenade, round cuff three inches deep, with slash flap on the sleeve six and a half inches long, two and a quarter inches wide at the points, and two inches wide at the narrowest point of the curve, four small buttons on the flaps, two on the hips, and four at regular distances along the skirt, and on the end of the skirt an embroidered shell and flame; coat to extend to within three inches of the bend of the knee, the skirt to be edged and lined with buff cloth.

For Company Officers.

Coat—same as for infantry, except the button.

Trousers—same color as the coat, with a stripe of buff cloth one half inch wide up each outward seam; leather strap to go under the boot; plain white for summer.

Vest—

Epaulettes—according to pattern.

Button—as at present adopted.

Cap—same as infantry.

Pompoon—yellow with red top.

UNIFORM OF THE ARMY.

Uniform, or Dress of the Army.

Boots—ankle.
Stock—black leather.
Sash—
Sword and Scabbard—according to pattern.
Waist-belt—white, with plate according to pattern.
Gloves—white.

Non-commissioned Officers, Musicians, & Privates.

Coat
Trousers } same as infantry, with the exception of the button.
Cap, &-c.

The *undress and fatigue* of officers and men of the Ordnance, will be the same as infantry, with the exception of the button, and the cord down the seam of the trousers, which will be of buff.

Shoulder Straps to be worn on the Undress Coat.

FOR A MAJOR GENERAL—strap of black velvet, four inches in length, and not more than one and a half inches in breadth, bordered with an embroidery of gold, one fourth of an inch wide, and three stars of five points each, within the embroidery.

FOR A BRIGADIER GENERAL—same as the above, with but two stars.

FOR A COLONEL—strap of the same size and color as the above, the border to be but one eighth of an inch wide, with an embroidered star in the centre of the strap, three crossed arrows embroidered on each end of the strap, the points towards the star; the border uniformly to correspond with the button of the corps; the star and feather of the outer arrows to be of silver where the border is of gold, and gold where it is of silver.

FOR A LIEUTENANT COLONEL—same as above, omitting the centre arrow at each end of the strap.

FOR A MAJOR—same as for a Colonel, omitting the two outer arrows at each end of the strap.

FOR A CAPTAIN—omitting the arrows and substituting two embroidered bars at each end of the strap, of the same width as the border, to be placed parallel to the ends of the strap,

UNIFORM OF THE ARMY.

Uniform, or Dress of the Army.

the distance between them and the border, equal to the width of the border; the bars to be of silver where the border is gold, and gold where it is silver.

FOR A FIRST LIEUTENANT—same as for a Captain, omitting one of the bars at each end of the strap.

FOR A SECOND LIEUTENANT—same as for a Captain, omitting the bars.

Captains to wear one strap on the right shoulder; Lieutenants one on the left.

Horse Furniture for Mounted Officers.

Housing for General Officers—of dark blue cloth, cut square, and to be worn over the saddle, with two rows of buff cloth, the outer row three fourths of an inch wide, the inner row two inches, with an embroidered star of five points on each side; the housing to be made so as partly to cover the horses haunches and forehands.
Bridle and Collar—buff leather.
Surcingle—yellow web.
Holsters—with black patent leather

Saddle Cloth for Mounted Regimental Officers.

FIELD OFFICERS AND COMMISSIONED STAFF—dark blue cloth, cut square, and made to wear under the saddle, with one stripe of cloth two inches wide, the stripe to correspond in color with the trimmings of the regiment.

COMPANY OFFICERS—same as above, except the stripe, which will be but one an a half inches wide.
Bridle—black leather; roses for the Cavalry, buff; for the Artillery, red; for the Ordnance, blue; for the Infantry, white.
Collar—black patent leather, with a metallic star of five points in front.
Surcingle—to correspond in color with the roses of the bridle.
Holsters—same as for general officers.
All metallic mountings, stirrups, bits, buckles, &c., of saddle and bridle, to correspond in color with the button of the corps.

UNIFORM OF THE ARMY.

Standards and Guidons.

Standards and Guidons for Cavalry.

Each regiment will have a-silken standard, and each troop a silken guidon. The standard will be the same as the national color, with the number and name of the regiment, in a scroll above the star. The flag of the standard to be two feet six inches wide, and two feet four inches on the lance.

The flag of the guidon to be made swallow-tailed, three feet from the lance to the end of the slit of the swallow-tail, and two feet on the lance. To be of blue silk, with a white star in the centre, and the letter of the troop (in red) within the star. The lance of the standards and guidons to be nine feet long, including the spear and ferule.

APPENDIX II
Navy of the Republic of Texas uniform regulations
facsimile reproduction

NAVAL GENERAL ORDER.

Navy Department,
13th March, 1839.

Hereafter, the Uniform Dress of the Officers of the Navy of the Republic of Texas shall be as hereinafter described, and to which all Officers are directed to conform.

NAVY UNIFORM.

CAPTAINS.

FULL DRESS. COAT of dark blue cloth lined with white, double breasted, with long lapels, the width of which is to be in proportion to the size of the coat, and cut with a swell; the lapels are to be buttoned back with nine buttons on each lapel. *Collar* to be lined with white, and embroidered in gold round the upper edges and sides with a rope and a broad pennant worked in gold, as long as the collar will admit, with a star of the same in the centre thereof, as per pattern. The *Cuffs* to have four buttons, and buttonholes worked in twist, and embroidered with live oak leaf and acorn, with a rope on the upper part, above the button, as per pattern. The *Pocket-flaps* to be embroidered the same as cuffs, the lower part and sides to have a rope and four buttons underneath. The *front* of the coat, on both sides, to be embroidered in gold with live oak leaves and acorns interspersed, commencing at the collar and running down the breast and skirts to the tail, the lower part of the breast rounded off to the skirt; one button on each hip, two under the middle of the folds and one at the bottom of each skirt; the pockets to be in the fold. Two gold epaulets, one on each shoulder.

VEST, white, single breasted, with as many small navy buttons as are worn on the breast of the coat. *Collar*, standing, coming to the edge of the breast and sloping in a line with it. *Breast*, straight, with pocket flaps, under each of which there shall be four small buttons.

BREECHES, white, with small navy buttons and gold or gilt kneebuckles—or plain white pantaloons over short boots, or with shoes and buckles.

UNDRESS. Coat in all respects the same as full dress, except the embroidery of live oak leaf and acorn from the collar along the breast and skirt to the tail, none of which shall be upon the undress. Or, a *double breasted Frock Coat*, lined with white, standing collar, and epaulets as in full dress.

VEST, either white or blue, same as in full dress.

PANTALOONS, plain blue, or, in warm weather, white; to be worn over half-boots, or shoes and stockings.

COMMANDERS.

FULL DRESS. Same as Captain, with the following exceptions: COAT to be without the embroidery of live oak leaf and acorn from the collar along the breast and skirts to the tail; and a long pennant embroidered upon the collar instead of the broad, and longer and less than half the width of that for Captain. Three buttons on the wrists and under the pocket-flaps, instead of four; one on the folds of the skirts, instead of two.

UNDRESS. COAT of dark blue cloth, frock, double breasted, standing collar, lined with white; buttons the same as in full dress. A stripe of gold lace and one epaulet on each shoulder, or stripes without epaulets; in other respects same as Captain.

LIEUTENANTS.

FULL DRESS. Same as Commander, with the following exceptions: No embroidery upon the cuffs. One epaulet, instead of two, and that to be worn upon the right shoulder. A speaking-trumpet embroidered upon the collar, instead of the long pennant, as per pattern.

UNDRESS. Same as Commander, except one epaulet instead of two.

PASSED MIDSHIPMEN.

FULL DRESS. Same as Lieutenant, with the following exceptions: *Collar* to be embroidered with a foul anchor and star of five points, (as all the embroidered stars in the naval uniform must be,) arranged as per pattern; no epaulets.

UNDRESS. Same as Lieutenant, except the epaulet.

MIDSHIPMEN.

FULL DRESS. COAT, dark blue cloth, lined with white, standing collar, single breasted; nine buttons on the right breast and short button holes on the left. Embroidery, a star upon the collar, and none on the pocket-flaps.

UNDRESS. Same as Passed Midshipman, except the star on the collar. They will also be entitled to wear a round jacket, single breasted, standing collar, and an anchor inserted thereon: buttons the same.

SURGEONS.

FULL DRESS. Same as Lieutenant, with the following exceptions: no epaulet; collar embroidered with the club of Æsculapius, and two stripes of gold lace upon the wrists of coat, one half inch wide each.

UNDRESS. Same as Lieutenant, except the epaulet; collar plain; stripes on the wrists same as full dress.

ASSISTANT SURGEONS.

Full Dress. Same as Surgeon, except one stripe of gold lace on the wrists, instead of two.

Undress. Same as Surgeon, except one stripe of gold lace on the wrists, instead of two.

PURSERS.

Full Dress. Same as Surgeon, except to substitute the cornucopia instead of the club of Æsculapius, and no stripes of lace upon the wrists.

Undress. Same as Surgeon, except the stripes of lace upon the wrists.

GUNNERS.

Full Dress. Same as Passed Midshipman, with the following exceptions: a great gun upon the collar, as per pattern, instead of the anchor and star, and no embroidery upon the pocket-flaps.

Undress. A round jacket, rolling collar, one small button upon each side of the collar, double breasted, nine buttons on each breast, and three upon each pocket-flap.

BOATSWAINS, CARPENTERS AND SAILMAKERS.

Full Dress. Same as Gunner, except the Boatswain to have a call embroidered upon the collar; Carpenter a broad-axe; Sailmaker a jib.

Undress. Same as Gunner.

CHAPLAINS.

Plain black coat, vest and pantaloons, to be worn over boots or shoes; or, black breeches and silk stockings with shoes. Coat to have three black covered buttons under the pocket-flaps and on the cuffs.

SCHOOLMASTERS AND CLERKS.

Plain blue cloth coat, single breasted, rolling collar, and made according to the fashion prevailing among the citizens at the time, with six navy buttons on each breast, one on each hip, and one at the bottom of the skirts.

CHIEF ENGINEER.

Same as Pursers, except the embroidery on the collar, which will be the lever-beam of the engine.

EPAULETS.

All officers entitled to wear epaulets are to wear gold lace straps on their shoulders three-quarters of an inch wide, to distinguish their rank when without epaulets. Epaulets are not to be worn, in foreign ports, with round hats, but with cocked hats or caps.

Captains are to wear two epaulets of gold, each with two rows of bullion; on each strap of the epaulet, an anchor and star in gold. Captains in command of squadrons, by order of the Secretary of the Navy, to have the star above the anchor of silver, during the time they are in actual command. The Senior Officer of the Navy, at all times, entitled to wear the silver star.

Commanders to wear two epaulets of gold, the same as Captains, except the ornaments on the straps.

Lieutenants to wear one epaulet of gold, plain, like the Commanders, on the right shoulder.

SWORDS

Are to be basket hilted cut and thrust; the blade not to exceed thirty inches in length, nor to be less than twenty-six, and to be slightly curved; in breadth not to exceed one inch and three tenths, nor less than one inch and two tenths; the gripe of those for Captains, Commanders, Lieutenants and other commissioned officers and Midshipmen to be white; of other officers entitled to wear swords, black; all to be yellow mounted, and basket hilts, with heads like the navy-buttons, and black leather scabbards. All Officers in full dress, or when wearing their epaulets on shore, are to wear swords, except Chaplains, Schoolmasters and Clerks.

Belts. Blue webbing for undress, white webbing for full dress; the clasp to bear the coat of arms of the republic.

Sword-knots. Blue and gold rope, with twelve gold bullions.

All Officers, when in undress, will be allowed to wear dirks, yellow mounted, except in foreign ports.

HATS.

All the Officers, except the Chaplains, Schoolmasters, Clerks, Boatswains, Gunners, Carpenters and Sailmakers, are to wear, in full dress, cocked hats, bound with black silk riband, to show one inch and a half on each side, with gold tassels formed with five gold and five blue bullions, and a black silk cockade with a plain gold or gilt star one inch in diameter in the centre.

Captains and Masters Commandants to wear, when in full dress, gold laced cocked hats, with a solid gold star one inch and a half in diameter on the cockade.

Officers, in undress, may wear blue cloth caps, with or without epaulets. Captains, Commanders and Lieutenants to wear a band of gold lace, one inch and a half wide, around their caps; the caps of all other officers to be plain.

STOCKS.

All officers are to wear stocks or cravats of black, the white of the shirt collar to be shewn above it, both in full and undress.

GENERAL LICENSE.

All Officers, when on board ship, may wear short blue jackets, with the number of buttons as designated for their respective coats; grey cloth, brown drilling, yellow or blue nankeen trowsers; also black or figured vests, of any colour except red.

All Officers will be permitted to wear, as undress, a citizen's blue cloth dress coat, cut after the fashion prevailing at the day, with the requisite number of buttons as in full dress, and the straps on the shoulders for those who are entitled to wear epaulets.

All Officers commanding by order from the Navy department to have a binding of gold cord around the bottom of the collar, the breasts and the skirts to the tail.

APPENDIX III
Republic of Texas Military Buttons

ARMY

1. Officers button. This is thought to be the earliest Texas military button. Star with beaded edge, with *REPUBLIC OF TEXAS* above, *ARMY* below. Gold (gilt) for staff officers, silver (white metal) for line officers. Large button for coat was 23mm, smaller button for vests and cuffs was 15mm. Scovill, 1836.

2. Staff. Star with *TEXAS* above. Gold. Large button 22mm, smaller one 15mm. Scovill, 1837.

3. Infantry. Star with an I in its center. *REPUBLIC OF TEXAS* around star. The First Regiment of Infantry was the elite unit of the army. Its strength was to be 560 men. This figure appears often in the contracts and invoices. Such uniform orders were undoubtedly for the First Regiment. Silver. Large size button was 19.5mm, smaller 14mm (the smaller version had only *TEXAS* above the star). Scovill, 1837.

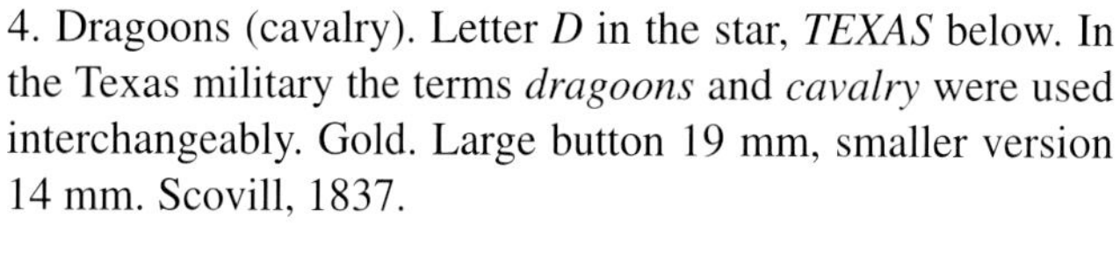
4. Dragoons (cavalry). Letter *D* in the star, *TEXAS* below. In the Texas military the terms *dragoons* and *cavalry* were used interchangeably. Gold. Large button 19 mm, smaller version 14 mm. Scovill, 1837.

5. Artillery. Star at top with *A* in it, crossed cannons below and *TEXAS* at bottom. Gold. Coat button 19mm, vest button 14mm. Scovill, 1837.

6. Light Artillery. Star at top with *LA* in it, crossed cannons below, *TEXAS* at bottom. Gold. Coat button 19mm, vest 14mm. Scovill, 1837. There may have been a similar button by Garanti of Paris, France.

7. Ordnance Staff. Flaming bomb at top, cannons below, *TEXAS* at bottom. Gold. Large button 23mm, smaller one 15mm. Scovill, 1837. Also another version possibly made in France.

NAVY

8. *REPUBLIC OF TEXAS* at top, star resembling a starfish just below, fouled anchor slanted to right below star. Gold. Coat button 21.5mm, vest and cuff button is smaller. Scovill, 1836.

9. *REPUBLIC OF TEXAS* at top, star just below has more conventional shape, below is a fouled anchor slanted to the right. Gold. Coat button 22mm, vest button is smaller. Made by Garanti, Paris, France, Date unknown.

10. *ESPERANZA* at top, large star below, straight up fouled anchor at bottom. Border is thinner at top and bottom. Gold. Coat button 23mm, vest button 15mm. Scovill, 1837.

11. Tiny star at top, vertical fouled anchor below, wreaths of oak and laurel leaves to left and right of enclosed words *TEXAS NAVY*, one on each side of anchor. Gold. Large button 23mm, small button 16mm. Scovill, 1840. One French-made version has been found, size 15mm, marked TW&W, Paris. This would be Trelon, Weldon & Weil, a large and well-known maker of military buttons.

MARINE CORPS

12. Banner with swallow tail at left end, hand holding sword at right end, banner encircles a tiny star with a vertical fouled anchor below the star. On the button banner at bottom is T-•-M-•-C. Gold. Large button is 21mm, small one 15mm. Scovill, ordered in 1839, delivered in 1840.

ACKNOWLEDGMENTS

All button photos were furnished by Gary Embrey of Longview, Texas, who has graciously shared his expertise on Texas military buttons with the author. Special thanks also to Chris Kneupper of the Brazosport Archaeological Society for sharing his knowledge of Texas Marine buttons.

SELECTED BIBLIOGRAPHY

PRIMARY SOURCES

Archival Collections and Exhibits

Archives and Library Division, Texas State Library. Army Correspondence and Papers. Navy Papers. Navy Yard and Ship Pursers Records. Quartermaster Records. Treasury Department and Auditors Records. Walke (Alfred) Journal 1840-1843.

Center for American History, University of Texas at Austin. Hefter (J.). Uniform sketches. Johns (Edward) Papers 1841-1843 and sketches. Prints and Photographs Collection.

Chapultepec Palace Museum, Mexico, D.F., Mexico. Texas flags and artifacts.

Rosenberg Library, Galveston Texas. Navy Papers and artifacts.

San Jacinto Monument, La Porte, Texas. Texas uniforms and research materials.

PUBLISHED PRIMARY MATERIALS

Books, Pamphlets and Bulletins

Bolleart, William. *William Bolleart's Texas.* Norman: University of Oklahoma Press, 1956.

Brown, Frank. *Annals of Travis County and the City of Austin.* 5 vols. Austin: typescript, Genealogy Room, Texas State Library [ca. 1875].

Delgado, Pedro. *Mexican Account of San Jacinto.* Austin: Institution for the Deaf and Dumb, 1878.

De Shields, James T. *Tall Men With Long Rifles: Set Down and Written Out by James T. De Shields as Told to Him by Creed Taylor, Captain During the Texas Revolution.* San Antonio: The Naylor Co., 1935.

Dewees, William B. *Letters from an Early Settler of Texas.* Louisville, printed by the New Albany Tribune, 1858.

Ehrenberg, H(erman). *Fahren und Schicksale Eines Deutschen in Texas.* Leipsig: Verlag von Otto Wigand, 1845. In English as *With Milam and Fannin: Adventures of a German Boy in Texas' Revolution.* Translated by Charlotte Churchill. Edited by Henry Smith. Forward by Herbert Gambrell. Dallas: Tardy Publishing Co., 1935.

Gaddy, Jerry J., comp. and ed. *Texas in Revolt: Contemporary Newspaper Account of the Texas Revolution.* Fort Collins, Colo.: Old Army Press, 1973.

Gray, Alfred G. [An Ex-Salt of the Texas Navy, pseud.] "Campeche Was Texas Navy's Last Engagement." *Port of Galveston Bicentennial Appointment Calendar and Compendium for 1976.* Galveston: Port of Galveston, 1976.

Gray, William Fairfax. *From Virginia to Texas, 1835: Diary of Colonel William Fairfax Gray, Giving Details of His Journey to Texas and Return in 1835-1836 and Second Journey to Texas in 1837.* 1909; reprint, Houston: Fletcher Young Publishing Co., 1965.

Green, Thomas Jefferson. *Journal of the Texian Expedition Against Mier, Subsequent Imprisonment of the Author, His Sufferings and Final Escape from the Castle of Perote With Reflections Upon the Present Political and Probable Future Relations of Texas, Mexico and the United States.* Austin: Steck Co., 1935 (reprint of 1845 edition).

Jenkins, John H., ed. *The Papers of the Texas Revolution, 1835-1836.* 10 vols. Austin: Presidial Press, 1973.

Jenkins, John Holland. *Recollections of Early Texas: The Memoirs of John Holland Jenkins.* Edited by John Holmes Jenkins III. Austin: University of Texas Press, 1958.

Johnson, Frank W. *A History of Texas and Texans.* 5 vols. Edited by Eugene C. Barker. Chicago and New York: American Historical Association, 1914.

Kendall, George Wilkins. *Narrative of an Expedition Across the Great Southwestern Prairies From Texas to Santa Fe*, 2 vols. London: David Bogue, 1855.

Maverick, George Maddison and Mary A. Maverick. *Memoirs of Mary Maverick.* San Antonio: Alamo Printing Company, 1921.

Peña, José Enrique de la. *With Santa Anna in Texas: a Personal Narrative of the Revolution*. Translated and edited by Carmen Perry. College Station: Texas A&M University Press, 1975.

Seguín, Juan N. *Personal Memoirs of John N. Seguín, from the Year 1834 to the Retreat of General Woll from the City of San Antonio, 1842*. San Antonio: Ledger Book and Job Office, 1858.

Sheridan, Francis C. *Galveston Island, or a Few Months off the Coast of Texas, the Journal of Francis C. Sheridan, 1839-1840*. Austin: University of Texas Press, 1954.

Smithwick, Noah. *The Evolution of a State; or, Recollections of old Texas Days*. Compiled by Nanna Smithwick Donaldson. 1900; reprint, Austin: University of Texas Press, 1983.

Stiff, Colonel Edward. *The Texas Emigrant, Being a Narration of the Adventures of the author in Texas, and a Description of the Soil, Climate, Productions, Minerals, Towns, Bays, Harbors, Rivers, Institutions and Manners and Customs of the Inhabitants of That Country: Together With the Principal Incidents of Fifteen Years, Revolution in Mexico: And Embracing a Condensed Statement of Interesting Events in Texas, From the First European Settlement in 1692, Down to the Year 1840*. Cincinnati. George Conclin, 1840. Waco: reprint by Texian Press, 1968.

Texas Navy in Yucatan, 1837, the. Translated by Richard Saiser. Edited and Annotated With an Introduction by Sandra L. Myres. Galveston: Rosenberg Library Bulletin N.S. vol. 6, No. 1, March, 1976.

Winfrey, Dorman and James M. Day. *The Texas Indian Papers*. 6 vols. Austin: Texas State Library, 1959-1960.

Newspapers

Austin *Texas Democrat*, Feb. 20, 1846.

Clarksville *Northern Standard*, 1846.

Houston *Telegraph and Texas Register*, 1846.

New Orleans *New Orleans Bee*, Vol. IX, No. 50, November 21, 1835.

New Orleans *Daily Picayune*, 1846.

SECONDARY MATERIALS

Books, Pamphlets and Bulletins

Abernathy, Francis E. *Observations and Reflections on Texas Folklore*. Austin: Encino Press, 1972.

Bancroft, Hubert Howe. *History of the North Mexican States and Texas*. 2 vols. San Francisco: A.L. Bancroft and Co., 1884.

Barker, Eugene C. *The Life of Stephen F. Austin, Founder of Texas, 1793-1836. A Chapter in the Westward Movement of the Anglo-American People*. 1925; reprint, Austin: Texas State Historical Association, 1949.

—. *Mexico and Texas 1821-1835*. Dallas: P.L. Turner Co., 1928.

Bate, Walter N. *General Sidney Sherman, Texas Soldier, Statesman, and Builder*. Waco: Texian Press, 1974.

Brown, John Henry. *History of Texas from 1685 to 1892*. 2 vols. St. Louis: L.E. Daniell, 1892-93.

Carter, Hodding. *Doomed Road of Empire*, New York: McGraw-Hill, 1963.

Chambers, William Morton. *Sketches in the Life of General Thomas Jefferson Chambers of Texas*. Galveston: Book and Job Office of *The Galveston News*, 1853.

Conner, John Edwin. *Flags of Texas*. Norman: Harlow Publishing Co., 1964.

Cox, Mamie Wynne. *The Romantic Flags of Texas*. Dallas: Banks Upshaw and Co., 1936.

De Shields, James T. *They Sat in High Places: the Presidents and Governors of Texas*. San Antonio: the Naylor Co., 1940.

Dienst, Alex. *The Navy of the Republic of Texas*. Temple, Texas; 2 vols. Source Texana Series, 1909. Fort Collins, Colo.: reprint, Old Army Press (n.d.).

Fehrenbach, T.R. *Lone Star: A History of Texas and the Texans*. New York: Macmillan Co., 1968.

Foote, Henry Stuart. *Texas and the Texans; or, Advance of the Anglo-Americans to the Southwest Including a History of Leading Events in Mexico, From the Conquest of Fernando Cortes to the Termination of the Texas Revolution*. 2 vols. Philadelphia: Thomas, Cowperthwait & Co., 1841.

Gambrell, Herbert Pickens. *Anson Jones: the Last President of Texas*. Austin: University of Texas Press, 1964.

Gilbert, Charles E., Jr. *A Concise History of Early Texas, as Told by its 30 Historic Flags*. Houston: Charles W. Parsons, Publisher, 1964 (reprinted 1971).

Hardin, Stephen L. *Texian Iliad, a Military History of the Texas Revolution, 1835-1836*. Austin: University of Texas Press, 1994.

Hefter, J(oseph). *The Army of the Republic of Texas*. Bellevue, Nebraska: Old Army Press (n.d.).

—. *The Navy of the Republic of Texas*. Bellevue, Nebraska: Old Army Press (n.d.).

Hill, Jim Dan. *The Texas Navy, in Forgotten Battles and Shirtsleeve Diplomacy*. Chicago: University of Chicago Press, 1937. Austin: facsimile reproduction of the original by State House Press, 1987.

Hogan, William R. *The Texas Republic: A Social and Economic History*. Norman: University of Oklahoma Press, 1946.

Hopewell, Clifford. *Sam Houston, Man of Destiny, a Biography*. Austin: Eakin Press, 1987.

Jackson, Jack. *Los Tejanos*. Stamford, Conn.: Fantagraphics Books, Inc., 1982.

—. *Recuerdan El Alamo, the True Story of Juan N. Seguín and His Fight for Texas Independence*. Berkley: Last Gasp, 1979.

James, Marquis. *The Raven, a Biography of Sam Houston.* Indianapolis: the Bobbs-Merrill Co., 1929.

Koury, Michael J. *Arms for Texas, A Study of the Weapons of the Republic of Texas.* Fort Collins, Colo.: Old Army Press, 1973.

Kredel, Fritz. *Soldiers of the American Army 1775-1945.* Chicago: Henry Regenery Co. (n.d.).

Lindheim, Milton. *The Republic of the Rio Grande.* Waco: W.M. Morrison, bookseller, 1964.

Loomis, Noel. *The Texan-Santa Fe Pioneers.* Norman: University of Oklahoma Press, 1958.

Lord, Walter. *A Time to Stand.* New York: Harper & Brothers, 1961.

Matthews, Jay A., Jr. *The Ten Battle Flags of the Texas Revolution.* Austin: Presidial Press, 1975.

Morton, Ohland. *Terán and Texas, A Chapter in Texas-Mexican Relations.* Austin: Texas State Historical Association, 1948.

Nance, Joseph Milton. *After San Jacinto: The Texas Mexican Frontier, 1836-1841.* Austin: University of Texas Press, 1963.

Nevin, David. *The Texans.* New York: Time-Life Books, 1975.

Oates, Stephen B., ed., *The Republic of Texas.* Palo Alto: American West Publishing Co. and the Texas State Historical Association, 1968.

O'Connor, Kathryn Stoner. *The Presidio La Bahia del Espiritu Santo de Zuniqa, 1721 to 1846.* Austin: Von Boeckmann-Jones Co., 1966.

Pierce, Gerald S. *Texas Under Arms: the Camps, Posts, Forts & Military Towns of the Republic of Texas, 1836-1846.* Austin: Encino Press, 1969.

Rosenfield, John, Jr. *Texas History Movies.* Illustrations by Jack Patton. Dallas: the Southwest Press, 1928.

Sanchez Lamego, Miguel A. *The Siege and Taking of the Alamo.* Some comments on the battle by J. Hefter. Translated by Consuelo Velasco. Santa Fe: Blue Feather Press for the Press of the Territorian, 1968.

Scott, Zelma. *A History of Coryell County.* Austin: Texas State Historical Association, 1965.

Shelton, Bob and Alvin Luckenbach. *Uniform Buttons of the Republic of Texas.* Dallas: (n.p.), 1986.

Siegel, Stanley. *A Political History of the Texas Republic, 1836-1845.* Austin: University of Texas Press, 1956.

Teja, Jesús de la. *A Revolution Remembered: The Memoirs and Selected Correspondence of Juan N. Seguín.* Austin: State House Press, 1991.

Texas Citizen Soldiers. Washington, Tex.: flyer by the Star of the Republic Museum (n.d.).

Steffen, Randy. *The Horse Soldier 1776-1943.* 3 vols. Norman: University of Oklahoma Press, 1978.

Under Texas Skies. Vol. 1, No. 10, Austin: Texas State Historical Association, March 1951.

Tolbert, Frank X. *The Day of San Jacinto.* New York: McGraw-Hill Book Co., 1959.

Waterhouse, Charles. *Marines and Others, the Paintings of Colonel Charles Waterhouse USMC ret.* Edison, N.J.: Sea Bag Productions, 1994.

Webb, Walter Prescott. ed., *The Handbook of Texas.* 2 vols. Austin: Texas State Historical Association, 1952.

Weems, John Edward and Jane Weems, *Dreams of Empire, A Human History of the Republic of Texas 1836-1846.* New York: Simon & Schuster, 1971.

Wells, Thomas Henderson. *Commodore Moore and the Texas Navy.* Austin: University of Texas Press, 1960.

Wharton, Clarence. *Remember Goliad.* Glorietta, N.M.: Rio Grande Press, 1931 and 1968.

Wilcox, R. Turner. *The Mode in Hats and Headdress.* New York: Charles Scribner's Sons, 1948.

Winfrey, Dorman et al. *Six Flags of Texas.* Waco: Texian Press, 1968.

Wortham, Louis J. *A History of Texas From Wilderness to Commonwealth.* 5 vols. Fort Worth: Wortham-Molyneaux Co., 1924.

Yoakum, Henderson. *History of Texas From Its First Settlement in 1685 to Its Annexation to the United States in 1846.* 2 vols. New York: J.S. Redfield, 1855.

Articles

Howren, Alleine. "Causes and origin of the Decree of April 6, 1830." *Southwestern Historical Quarterly*, Vol. LXXXIII, No. 4, April 1980, 378-422.

Jenkins, William H. "The Red Rovers of Alabama." *The Alabama Review*, April 1965, 106-110.

Marshall, Bruce. "Commodore Edwin Moore: Texas Navy." Cover art by Bruce Marshall. *Military History of Texas and the Southwest*, Vol. X, No. 2, 1972.

Moore, Marc A. "Marines of the Texas Republic." Cover art and illustrations by Bruce Marshall. *Marine Corps Gazette*, Vol. 62, No. 8, August 1978.

Scarborough, Jewel Davis. "The Georgia Battalion in the Texas Revolution, a Critical Study." *The Southwestern Historical Quarterly*, Vol. LXIII, 1960, 511-532.

Newspapers

"Mexican officials lose Alamo flag." *Austin American-Statesman*, December 19, 1994.

THESES AND DISERTATIONS

Adams, Allen E. "The Leaders of the Volunteer Grays: the life of William G. Cooke, 1808-1847." M.A. thesis, Southwest Texas Teachers College, 1939.

Friend, Lorena B. "The Life of Thomas Jefferson Chambers." M.A. thesis, University of Texas at Austin, 1928.

UNIFORM PLATES

Hefter, J(oseph). *Uniforms of the Militia of Coahuila y Texas,* Military History Institute. Mexico, D.F., Mexico: (n.d.).

—. and Ed Milligan. *Texas Republic Navy, 1839.* Plate No. 395, Military Uniforms in America. The Company of Military Historians. Westport, Conn.: Winter 1973.

Jones, Tom. *Alabama Red Rovers, 1835.* Plate No. 502, Military Uniforms in America. The Company of Military Historians. Westport, Conn.: 1980.

—. *The First Texas Rangers, 1823.* Plate No. 457, Military Uniforms in America. The Company of Military Historians. Westport, Conn.: 1977.

—. Elting, John R. *The Alamo Garrison, March 1836.* Plate No. 386, Military Uniforms in America. The Company of Military Historians. Westport, Conn.: 1973.

Uniforms of the United States Navy 1776-1898. Naval History Division, Office of the Chief of Naval Operations. Washington, D.C.: 1966.

INTERVIEWS AND PERSONAL COMMUNICATIONS

Albert, A.H. (Dewey). Button collector , Highstown, N.J. to author February 8 and April 9, 1977 regarding Texas military buttons. Letters in author's possession.

Biffle, Kent. Texana columnist, *The Dallas Morning News.* Interview by author, November 14, 1994.

Butler, Pierce. Nashville, Tenn. Two telephone interviews by author, November 2 and 9, 1995, regarding painting in his possession by John Mix Stanley depicting Tehuacana Creek conference with Indians in 1843.

Elting, Colonel John R., USA (Ret.) Texana expert, Company of Military Historians, to author, March 25, 1995. Letter in author's possession.

Embrey, Gary. Longview, Texas. Collector of Texas military buttons to author June 13, August 3, September 7, 1995. Letters and enclosures in author's possession.

Green, Michael. Austin, Texas. Former Reference Archivist, Texas State Archives, former Assistant Editor, *Military History of Texas and the Southwest.* Telephone interviews by author September 24, 1994, January 5, 1995, February 2, 1995.

Hefter, Joseph. Mexico, D.F., Mexico and Cuernavaca, Mexico. Artist and military historian. Interview by author in Mexico City September 17, 1972. Letter to author December 13, 1972 in author's possession.

Keller, Marisa. Acting Archivist, the Corcoran Museum of Art/School of Art, Washington, D.C. to author July 11, 1994. Letter and enclosures in author's possession.

Kneupper, Chris. Brazosport, Texas. Brazosport Archaeological Society to author (n.d.). Note and enclosures in author's possession.

Koury, Michael J. Fort Collins, Colo. editor and publisher, Old Army Press. Telephone interview by author, February 3, 1998.

Matthews, Brigadier General Jay A., Jr. editor and publisher, Presidial Press, editor emeritus, *Military History of the West* (formerly *Military History of Texas and the Southwest*) Numerous telephone interviews by author, 1994, 1995, 1996.

Miller, Edward. San Antonio. High school history teacher with special interest and research concerning the New Orleans Grays to author May 19, 1995. Letter and enclosures in author's possession.

Milligan, Edward. Alexandria, VA. Texana specialist, Company of Military Historians to author June 17, 1994. Letter in author's possession.

Nesmith, Sam. San Antonio. Former curator of the Alamo and former researcher for the University of Texas Institute of Texan Cultures at San Antonio. Telephone interview by author August 25, 1995.

Roark, Garland. Nacogdoches, Texas. Novelist, author of *Star in the Rigging* about the Texas Navy, to author April 13,1973. Letter in author's possession.

Smith-Christmas, Kenneth L. Curator of Material History, Museums Branch, Marine Corps History and Museums Division, Quantico, Va., to author April 17, 1995. Letter in author's possession. Telephone interview by author [ca. May 12, 1995].

NOTES

NOTES

NOTES